CARTOGRAPHIC TREASURES
OF THE
NEWBERRY LIBRARY

THE NEWBERRY LIBRARY
OCTOBER 10, 2001 – JANUARY 19, 2002

ISBN 0–911028–71–4

Curators: James R. Akerman and Robert W. Karrow, Jr.
Exhibit Manager: Riva Feshbach
Exhibit Designer: Pat Chester
Conservator: Susan Russick
Photographer: Catherine Gass
Catalogue Editor: Sara Austin
Catalogue Designer: Aaron Shimer
Typeset by: Julie Breslin
Printed in the United States of America by IPP Lithocolor.

Dymaxion Airocean World (item 19) reprinted by permission of the Buckminster Fuller Institute.
Esso War Map (item 44) reprinted by permission of Langenscheidt Publishing Group.

Cartographic Treasures of the Newberry Library is made possible by the generous sponsorship of LaSalle Bank and Rand McNally.
Additional support was provided by Roger and Julie Baskes, Art and Jan Holzheimer, Mr. and Mrs. Arthur L. Kelly, Mary
Ann and Barry MacLean, Andrew McNally III, Jossy and Ken Nebenzahl, and Mr. and Mrs. Rudy L. Ruggles, Jr.

The Newberry Library, 60 West Walton Street, Chicago, IL 60610
www.newberry.org

CONTENTS

PREFACE

When the International Map Collectors' Society invited the Newberry Library to be one of two institutional hosts for its 20th International Symposium in October 2001, we instantly resolved to bring out the "good china" for our honored guests. The fruit of that resolve is *Cartographic Treasures of the Newberry Library,* the most comprehensive display of the Newberry's approximately 300,000 maps that the Library has ever mounted. Both the exhibit and the occasion that inspired it deserve a suitable memorial.

We hope that this catalogue will itself become a prized souvenir for those attending IMCoS 2001, which the Newberry Library has organized in close collaboration with the staff of the American Geographical Society Collection in Milwaukee, the Adler Planetarium, the Chicago Map Society, and the Map Society of Wisconsin. *Cartographic Treasures* also honors the occasion of the 14th Kenneth Nebenzahl, Jr., Lectures in the History of Cartography to be held October 11–13, 2001. This public lecture series, organized by the Newberry's Hermon Dunlap Smith Center for the History of Cartography, has been supported by the generosity of Mr. and Mrs. Kenneth Nebenzahl since 1966. The 2001 Nebenzahl Lectures, entitled *A Taste for Maps: Commerce and Cartography in Early Modern Europe,* concern the history of the business of mapmaking in France, Britain, Italy, the Netherlands, and Germany.

Exhibits and exhibit catalogues are like weddings. They require a great deal of planning; their success relies on the good will and talents of many people; and they require a substantial financial investment. As curators, we were indeed fortunate to be able to rely on the help of so many talented individuals. Riva Feshbach, the Newberry's exhibits manager, guided the assembly of the exhibit with her usual efficiency and patience, bending here and there to our whims, but keeping us on track and on budget. Pat Chester produced the elegant design for the show. Newberry director of conservation Susan Russick made it possible for us to include many large and fragile treasures in unusual formats. She and her conservation staff prepared all the exhibits for display, mending them where needed, with their usual skill. Sara Austin, assistant director of research and education for the Newberry, served heroically as managing editor for the catalogue, relieving the curators of most of the burden of bringing the catalogue to press and editing our text with insight and good judgment. Additional editorial assistance was provided by Sarah Fenton and Meg Moss and Associates, Inc. Brigid Murphy, the Newberry's director of communications, provided valuable guidance in the development of the catalogue concept and oversaw production of the catalogue, exhibit poster, and related publicity, to which Aaron Shimer lent his considerable talent as graphic designer. Staff photographer Catherine Gass nearly stood on her head to capture images of treasures running the gamut from wall- to pocket-sized. Jodi Morrison served as registrar for the exhibit, sorting out our often vaguely expressed desires. Susan Hanf, administrative assistant of the Smith Center for the History of Cartography, helped out with countless tasks, both large and small, maintaining the sanity of all involved.

For more than a century now, map collecting and map scholarship at the Newberry Library have grown and flourished through the stewardship and enthusiastic support of our many local friends. *Cartographic Treasures of the Newberry Library* continues this tradition. We are grateful for the generous sponsorship of LaSalle Bank and Rand McNally for the exhibit. We are honored also to have received financial support from Roger and Julie Baskes, Art and Jan Holzheimer, Mr. and Mrs. Arthur L. Kelly, Mary Ann and Barry MacLean, Andrew McNally III, Jossy and Ken Nebenzahl, and Mr. and Mrs. Rudy L. Ruggles, Jr. We are particularly grateful for the leadership of Mr. Baskes and Mr. Nebenzahl. *Cartographic Treasures* is but the most recent product of their tireless support of map collecting and scholarship at the Newberry.

INTRODUCTION

What makes a map a treasure? A simple question, but for the curators and readers who have come to know the Newberry Library's map collection well over many years, not one with a simple answer. "Treasure" usually refers to something of great rarity or value. But while the odds and ends a child stows away in a private "treasure box" may be worth little in the marketplace, they are valuable beyond measure to the child. Creating this exhibit and catalogue has been an extended meditation on what we treasure about the Newberry's maps.

The approximately 300,000 maps in the Newberry Library include many that are rare, beautiful, and valuable in the conventional sense. Some of the finest examples of these find pride of place in *Cartographic Treasures of the Newberry Library*. Of the centuries-old printed maps reproduced here, for example, some survive in but a few copies worldwide, while others—such as Kaspar Vopel's 1597 map of Europe (item 10)—can be found only at the Newberry. The exhibit also includes the Newberry's oldest cartographic document: Gregorio Dati's *La Spera* (item 1), a hand-illuminated geography and astronomy textbook dating from about 1425. And the charts in the magnificent manuscript world atlas drawn by the Portuguese chartmaker Sebastião Lopes around 1565 (item 20) have few rivals for their beauty.

Yet the readers who have visited the Newberry's Rand McNally Map Reading Room over the years—teachers, scholars, and map enthusiasts alike—have shown us that people prize maps for various reasons. As with so many other things, maps' costliness, rarity, or beauty may catch the eye, but their deeper worth lies elsewhere. The value of carto-graphic documents resides in their ability to speak about the past, to bring to life the people who made or used them, and to animate the landscapes they depict and the cultures that produced them. The maps selected for this exhibit also exemplify this broader understanding of "treasure." Whether rare or common, all appear here because they evoke so well the spirit of their age. Thus, the warring colors of John Mitchell's wall map of North America in 1755 (item 39) conjure up the heated colonial rivalry between Britain and France in North America. W. H. Holmes's view of "The Grand Cañon at the Foot of Toroweap—Looking East" (item 33) embodies the wonder American explorers and surveyors felt when they encountered the sublime landscapes of the West. Even such mass-produced maps as Rand McNally's Auto Trails map of the American Southwest in 1923 and Esso's war map of 1942 (items 57 and 44) are prized precisely because, through their ubiquity, they deeply influenced peoples' understanding of their world.

The limited size of the exhibit posed the greatest challenge to showcasing the Newberry's cartographic treasures. According to our broad definition, virtually all of the Library's 300,000 maps are treasures, but there was space for fewer than eighty of them. Some obvious inclusions quickly rose to the top of our list. Beyond that, we endeavored to summarize the biography of the map collection itself, how it grew, and what its great strengths are. The historic core collections are all represented, as are noteworthy recent purchases and gifts that have kept the Newberry's map holdings responsive to readers' changing needs and new scholarly trends. Those who know the Newberry well might regret the omission of their personal favorites. As we prepared this catalogue, we too were daily tempted to make additions. In resisting those temptations, we fell back on the reassurance that map collections are living and constantly evolving: they reveal their value by study and use, rather than by distant admiration. As future generations sift and resift the Newberry's collections, we trust that they will find new stories to tell and new maps to treasure.

THE COLLECTIONS

The Newberry Library's map collection is, in many respects, a microcosm of the Library's collections as a whole, mirroring the Library's greatest strengths. Maps produced by Western mapmakers of the European and American continents predominate, although all regions of the earth are well represented. The Newberry is particularly rich in Renaissance and early modern cartography; materials related to the Transatlantic encounter between Europe and America; the mapping of the exploration, non-native settlement, and exploitation of the American West; local cartography, particularly of the Midwest; and transportation cartography. Like the Library at large, the Newberry's map collection is essentially a mosaic of many individual collections acquired by bequest, gift, or purchase. The sum of these parts is testimony to the enduring support of Chicago's generous and visionary cartographic community.

The historic core of the Newberry's map collection rests largely on the magnificent Edward E. Ayer Collection of Americana. An early trustee of the Library, Edward Ayer had an abiding fascination with both the European exploration and colonization of America and the history and culture of American Indians. He donated his vast collections of books, manuscripts, paintings, prints, and maps relating to these subjects to the Newberry in stages from 1895 to 1927. His contribution to the Library's cartographic holdings includes 2,000 separately printed maps, 500 atlases, and 300 manuscript maps.

But mere numbers cannot express the breadth and rarity of the Ayer maps and atlases, many of which are unobtainable today. The catalogue of early modern European geographical works and atlases in his collection reads like an honor roll of fabled cartographers, including Bordone, Münster, Apian, Honter, Ortelius, Mercator, Hondius and Jansson, Blaeu, Speed, Visscher, de Wit, Jaillot, de Fer, Homann, Covens and Mortier, Robert de Vaugondy, Jeffreys, and many others. Among the many highlights are 22 portolan charts and atlases dating from the mid-fifteenth through the eighteenth century and a collection of editions of Claudius Ptolemy's *Geographia* assembled by the London dealer Henry Stevens. Still other treasures document the first centuries of contact between native Americans and newcomers from the Eastern Hemisphere. Among these are several rare sixteenth-century examples of cartography blending European and Mesoamerican styles (see items 60 and 67), and the *Cartes Marines,* an anonymous one-of-kind atlas documenting the extent and ambitions of French colonial power in the early eighteenth century (item 69).

Ayer's gift was complemented by a steadily growing corpus of maps in the Library's general collections, including many maps and atlases bought or donated by friends of the Library since the 1880s. The general collections include many intriguing examples of the development of cartography throughout the modern period (see items 16, 22, and 27); the Library's holdings of nineteenth- and early twentieth-century general and school atlases are particularly strong. Here, too, is an ample collection of American local geography and cartography, with a superb list of county land ownership atlases (see item 76). This cornerstone of the Newberry's renowned local history collection continues to expand thanks to the Newberry's commitment to family history and genealogy. The general collections also feature many railroad and tourist maps and cartography appearing in popular magazines, journals, and government documents, along with a rich and varied collection of maps of local Chicago interest (see item 66).

Since the Second World War, the Newberry community's awareness of the value of its cartographic holdings sharpened as the history of cartography took shape as a field in its own right. In response, under the leadership of President and Librarian Lawrence W. ("Bill") Towner (1962–86), the Library forged its role as a center for cartography through a series of institutional investments to encourage research. In 1966, the Library inaugurated the Kenneth Nebenzahl, Jr., Lectures in the History of Cartography, dedicated to the study of new topics in the field. This series has since launched nearly a dozen pathbreaking books in map history and continues to influence the course of the field. By creating the position of curator of maps in 1969, the Library ensured that map acquisitions were coordinated and based upon sound bibliographical research and scholarship. In 1972, the Library established the Hermon Dunlap Smith Center for the History of Cartography, the first research center dedicated specifically to map history anywhere in the world. The Smith Center's research fellowships, summer institutes for teachers and scholars, workshops, exhibits, and conferences have served thousands of scholars worldwide directly, and countless more through their publications and its own. The Smith Center's name honors Hermon Dunlap ("Dutch") Smith, like Ayer before him a president of the Library's board of trustees. Working closely with President Towner, he championed these developments as well as several major map acquisitions in the 1960s and 1970s. Smith established an endowment that supports in part the work of the Center and map curatorial staff and funds purchases of cartographic reference materials. Like that of many of the map collection's benefactors, Smith's enthusiasm for cartography rested in part on his own passion for collecting historical and geographical materials. Smith bequeathed to the Library his fine collection of Great Lakes history and geography, including many important maps of the region.

Led by Towner, Smith, and several trustees who shared their interest in cartography, the Newberry embarked upon an aggressive period of map acquisition that assured its reputation as one of the premier repositories of historic maps in the world. This expansion of the collections gathered momentum from the 1964 bequest by Everett D. Graff, another board president with a particular passion for the American West. Everett Graff was renowned for his judicious and well-balanced selection of materials, both rare and common. The significance and breadth of the Graff Collection's 5,000 volumes and maps are such that the published catalogue of the collection now serves as a standard reference for western Americana. The Graff maps, most of which appear in books and pamphlets, are a treasure trove—an appropriate cliché here—documenting the exploration and survey of the West, the development of its trails and railroads, its economic exploitation, the geography of Indian-white relationships, and immigration and settlement. Many are ephemeral maps that would have been lost were it not for Graff's wandering collector's eye. The 1862 map of the mining district around Central City, Colorado (item 74), for example, survives in only two known copies.

In the years to come, a series of major acquisitions deepened the Newberry's holdings in Renaissance and eighteenth-century maps of Europe and its colonies. In 1967, the Novacco Collection instantly established the Library's reputation as a repository of printed Italian works from the sixteenth and seventeenth centuries (see items 5, 7, 13, and 59). The 1,600 German, French, Dutch, and English printed maps in the Sack Collection continued the European chronicle for the seventeenth and eighteenth centuries (see item 26). The Visscher Collection added 124 maps published by Claes Janszoon Visscher, the talented seventeenth-century Dutch cartographer (see item 38). Two fine collections of military plans joined these. One, dealing largely with battles of the American Revolution (see item 41), supported research on the *Atlas of the American Revolution* (University of Chicago Press, 1974) compiled by Library staff and resident scholars. The other, comprising 189 plans relating to the European dynastic wars of the eighteenth century (see item 40), provides a superb—but still largely untapped—record of battle reportage directed at both military authorities and civilians.

Chicago's position as a national transportation hub is reflected in the strength of the Newberry's holdings of railroad and road maps. The immediate post–World War II period brought to the Library the archives of two Chicago-based railroads, the Illinois Central and the Chicago, Burlington, and Quincy. The CB&Q archives are particularly rich in the mapping generated by that railroad and its close associates promoting settlement and tourism along rail routes (see item 77). In 1989, the Newberry acquired the Rand McNally Collection, comprising the Chicago-based company's huge output of rail, road, and airline maps (see items 56 and 57), as well as its extensive list of reference and school atlases, guidebooks and geographies. In the last decade, the Library has sought to expand its holdings of twentieth-century road maps, establishing the Newberry as a leading repository of these valuable documents of American automobile culture. This effort bore fruit in 1998 when the American Map Company donated the archives of the General Drafting Company, another major publisher of twentieth-century road and tourist maps (see item 44).

Surprisingly, many of the maps that draw researchers to the Newberry year after year are not even originals. Rather, they form part of the Library's comprehensive holdings of cartographic facsimiles and reproductions, among them the compilations of photostats in the Karpinski and Ayer Collections. These photographic reproductions of 1,300 manuscript maps illustrating American colonial history were culled from widely scattered European archives. They are therefore much treasured by the Newberry's readers, since they allow scholars to consult maps from libraries around the world without leaving our reading rooms. These are truly "infinite riches in a little room." Some of these facsimiles are themselves rarities, or they reproduce maps that have since been lost to the ravages of time and war.

This brief sketch of the Newberry's first one hundred years of map collecting cannot include all the people and institutions that helped to amass its treasures, nor is this story at an end. The Library's map collections continue to receive valuable support from its trustees, its benefactors, and members of the local map-collecting community who share our enthusiasm for maps, geography, and history. Some have contributed treasures from their personal collections. Others have offered valuable and timely financial support that made possible major individual acquisitions. Still others have established endowed funds specifically dedicated to the acquisition of cartography. In all its forms, their generosity has allowed the curator of maps to exercise discretion in purchasing new items that fill significant gaps in the Newberry's holdings, that further develop collection strengths, and that keep the Library in the forefront of map collecting and scholarship. For example, Edmond Halley's map charting the phenomenon of magnetic declination (item 49), represents a crucial step in the evolution of scientific cartography that surely belonged in the Newberry, but we acquired this landmark only in 1986. We especially treasure Taylor and Skinner's 1776 road atlas of Scotland (item 50), a recent addition to our holdings of transportation cartography, because its original soft binding, suitable for eighteenth-century travel, has been preserved.

The set of unusual "typographic" maps published by Wilhelm and Georg Haas acquired in 1991 (see item 28) joined others already in the collection documenting the evolution of map production technology. These additions continue to shape a century-long relationship among collecting, scholarship, and teaching that typifies cartography at the Newberry.

EXHIBIT THEMES

The maps in *Cartographic Treasures of the Newberry Library* date from circa 1425 to 1954 and represent a wide variety of cartographic styles and purposes. The exhibition could have been organized chronologically, so its viewers and readers of this catalogue might see how cartographic styles, mapmaking technology, and the nature of geographical knowledge have evolved over time. Or we might have arranged these maps geographically, to demonstrate the in-depth coverage the Newberry can muster for virtually any region of the world. But the documents themselves reject any attempt to reduce the history of cartography into a single narrative. No one map can ever epitomize the mapmaking of its time or fully represent a region of the world. And in any event, the most compelling questions maps raise are not the "wheres" or "whens," but the "whys" and "hows": Why do people make the maps they do? How do they use maps to understand and shape their world? Looking over the 500-year span of cartographic history represented in the exhibit, we were struck by the timelessness of certain map uses. So, rather than emphasize change, we thought to emphasize the uses of cartography that were as relevant in early modern Europe and Mesoamerica as they were in the twentieth century.

Our survey begins with "Grasping the World," a selection of maps and atlases that grapple with the most basic of geographical questions: What does the world look like? This section showcases world and continental maps, as well as atlases and educational tools, that reflect westerners' changing sense of the outlines of the continents and oceans over time. Hence, we trace an evolution from the geographically limited view of the world Renaissance Europeans inherited from the ancient geography of Claudius Ptolemy (items 2 and 3); to the expansive baroque vision of the bloated atlases and gigantic wall maps of the seventeenth century (see item 12); to the scientifically and imperially driven exploratory maps of the nineteenth century (see item 15). We also wished to illustrate the diversity of forms cartographers used to help people understand that outline. The atlases in "Grasping the World" show how their makers have struggled with the problem of digesting a world of maps and organizing them into useful reference works. A selection of pedagogical maps, including a fifteenth-century textbook (item 1), nineteenth-century flashcards (item 14), and a jigsaw puzzle (item 17) represents various attempts to make geography entertaining and comprehensible to young scholars. This section, dedicated to the problem of reducing the earth to comprehensible form, concludes appropriately with a globe that collapses (item 18) and Buckminster Fuller's world map that can be assembled into a rough approximation of a globe (item 19).

The next section, "Inventing the Nation," ponders the idea that nations are not natural divisions among human beings, but are created by the human mind and human action. The maps selected here have all played some role in inventing nations—either as tools of government; or as the media of the political debates, news, and ideas that fashioned national identities and enmities. Three products of innovative systematic surveys of European countries in this section exemplify attempts by European monarchs and their ministers to gain better understanding—and therefore control—of their subject territory: Apian's map of Bavaria (item 21), Saxton's *Atlas of England and Wales* (item 22), and Cassini's atlas of French survey triangles (item 27). Other items explore the role of mapping in shaping a national geographical consciousness. Sebastião Lopes's magnificent atlas of 1565 (item 20), for example, reconstructs a Portuguese maritime journey around the world at the height of Portugal's maritime prestige and power. John Farrar's map of early Virginia (item 25) illustrates the wishful geographical thinking and commercial mentality that drove early European colonization of America. Pál Teleki's ethnographical map of Hungary in 1924 (item 34) demonstrates how maps can engage and influence diplomatic and public opinion during international disputes.

Our third section, "Contesting Places," showcases maps that report on the course and consequences of human conflict. Throughout modern history, warfare has been a powerful stimulus for mapping activity and innovation. This partly reflects military engineers' and strategic planners' need for geographic information. Alexander de Groote's 1617 treatise on fortification methods (item 36), for example, illustrates the reliance of military architects and ordnance experts on cartography to plan attacks on and defense of fortresses and cities. Most of the maps here, however, were intended to inform and engage a wider public in the events of the day. A colorful map published by the oil company Esso during the Second World War (item 44) is perhaps the baldest example of wartime propaganda among our war maps. Published shortly after the American entry into the war, it touted the wartime uses of petroleum products, while providing a geographic context civilians could use to better follow military operations as they unfolded. Claes Janszoon Visscher's narrative plan commemorating the successful Dutch siege of the Spanish garrison in Breda in 1637 (item 38) is different in tone and technical execution, but is no less patriotic than the Esso map.

"Conquering Distances" focuses on maps dedicated to movement, navigation, and conquest of space. We begin with several maps illustrating the early history of sea charts. One of the oldest Newberry maps is here, a fine example of a portolan chart of the Mediterranean Sea drawn by Petrus Roselli in 1456 (item 45). The elegantly simple construction of this chart may be contrasted with two charts that follow (items 47 and 49), dedicated to technical improvements in maritime navigation, and with the truly massive eighteenth-century survey of the Atlantic coast of North America by Joseph F. W. DesBarres (item 51). The maps that follow trace the refinement of interior communication in Europe and America from the seventeenth through the twentieth centuries and capture the fascination with innovations in overland transportation technology. Hence, the elaborate 1697 map of the French Canal de Languedoc (item 48) represents the dawn of the era of canal building. The 1866 Dodge map of the future Union Pacific Railroad (item 55) brings to life American excitement about the possibilities of the railroad in the development of the West. And Rand McNally's unique Photo-Auto Guide to the road from Chicago to Milwaukee (item 56) exemplifies the beginning of the American infatuation with the automobile. In fact, the items in this section should not be seen merely as navigational tools, but as reflections of the human appetite for movement and exploration.

The fifth section of the exhibit, "Celebrating the City," demonstrates how maps and views capture the essence of urban geography, architecture, and social life. We begin with two nearly contemporary but strikingly different images of Venice. Erhard Reeuwich (item 58) rendered the city as if from a watercraft, as indeed most travelers to the city would have viewed it. In 1500, Jacopo de Barbari (item 59) offered the technically impossible view of the city from the air. Cartographically, the advantages of this perspective are obvious. It allows one to see at a glance the entirety of the city to get a real sense of its geographical layout—something we can do easily from an airplane today, but that European viewmakers were just learning to do in their imagination. By the later sixteenth century, "bird's-eye" views of cities were commonplace, mainstays of popular published collections such as Braun and Hogenberg's *Civitates orbis terrarum* (item 61) and the townbooks published by the Dutch atlasmakers Jansson (item 62) and Blaeu. Military officials and planners preferred maps that rendered the city planimetrically, as if the viewer were directly overhead every point shown on the map. True plans, such as the 1776 depiction of Newport by Charles Blaskowitz (item 64) and the siege plan of Breda (item 38), map every part of the city at the same scale and thus allow for accurate calculation of distances and areas. But what they gain in accuracy, true plans lose in visual impression. For this reason, pictorial depictions of the urban landscape retained their popularity among those who celebrated the city in the nineteenth and twentieth centuries (see items 65 and 66).

A similar tension between the need for geometric accuracy and precision on the one hand and the greater legibility of more informal and pictorial styles of cartography on the other appears in "Plotting the Countryside," a selection of rural maps and views. Where there was a desire to subdivide land carefully, to calculate its economic potential, or to plan for its settlement, a measured plan or plot of the landscape was required. Examples of this sort of rural map include Jan Spruytenburgh's 1734 map of a typical Dutch polderland (item 68), Richard Richardson's 1758 *Plan of Miss Hall's Estate* (item 71), Whitten's 1820 map of the district of Spartanburg, South Carolina (item 73), and Pratt and Buell's large map of the gold claims and mines of Gilpin County, Colorado, in 1862 (item 74). But some of these maps also use pictorial elements to give an impression of the ambiance of the landscape they depict. The planimetric and the pictorial are juxtaposed in the same image on the map of Jonathan Miller's southern Illinois farm from 1874 (item 76). The gridded plan gives us a sense of the economy of the Miller farm, down to the number of acres devoted to individual agricultural uses; while the accompanying view gives us an impression of the place as Miller would like us to see it—prosperous, peaceful, well kept, and in tune with the latest architectural styles.

Our list of these timeless contexts of mapmaking is by no means an exhaustive survey of map use, nor does it illustrate each theme to its fullest potential—each really merits its own

fully developed exhibit. But we hope *Cartographic Treasures* offers its viewers and readers a sense of the territory. Wherever possible, we have supplied at the bottom of each explanatory caption a brief list of references that may be consulted to learn more about an exhibit item or its historical context. We have gathered these and other essential works on the Newberry's collections and on the themes of the exhibit into a bibliography to be found at the end of this book. We have also supplied the call number for each map or atlas in the exhibit, so that you may find it easily when you next visit. We hope *Cartographic Treasures of the Newberry Library* will inspire you to do so.

JAMES AKERMAN
Director, Hermon Dunlap Smith Center for the History of Cartography

GRASPING the WORLD

1 GREGORIO DATI, *La Spera* (FLORENCE, CA. 1425). [AYER MS MAP 1]

Fifteenth-century Florentine silk merchant Gregorio Dati's business took him to various ports in the Mediterranean and possibly western Europe as well. Dati likely intended "The Sphere"—his account, in verse, of cosmography, astrology, and geography—to be an elementary textbook for students seeking to grasp these difficult subjects. *La Spera* includes a T-in-O world map, so called because its schematic division of the known world into three parts (Europe, Asia, and Africa) looks like a "T" inside an "O." The section on the geography of the Mediterranean basin, shown here, features snippets of portolan charts as marginal illustrations. The geographical survey is limited to ports of the southern Mediterranean and Black Seas. Dati undoubtedly would have completed the Mediterranean circuit and perhaps even extended it into the Atlantic had he not died in 1435.

Reference: COOK.

2 CLAUDIUS PTOLEMY, [WORLD MAP], IN HIS *Cosmographia* (ULM: LEONHART HOLLE, 1482). [AYER *F6 P9 1482A]
See color plate on page 84.

AND

3 CLAUDIUS PTOLEMY, "TABULA OCTAVA DE EUROPA," IN FRANCESCO BERLINGHIERI AND CLAUDIUS PTOLEMY, *Geographia* (FLORENCE: NICOLAUS TODESCO, [1482]). [AYER *F6 P9 B5 1480B]

Translated into Latin in the early fifteenth century, the *Geographia* of second-century Greek scholar Claudius Ptolemy profoundly shaped Renaissance Europeans' understanding of their world. Ptolemy included detailed instructions for making maps and popularized the grid of latitude and longitude lines still basic to mapmaking today. The 27 maps usually attached to Renaissance editions of Ptolemy were based on his own tables of geographical coordinates computed for hundreds of ancient places. They offered Renaissance readers a far more comprehensive inventory of the geography of the Eastern Hemisphere than was available in other fifteenth-century sources.

The Newberry's Ayer Collection holds all but one of the early printed editions of the *Geographia*. The woodcut "world" map in the Ulm edition of 1482, with the variant title *Cosmographia,* shows only the quarter of the earth known to ancient scholars, with the addition of Scandinavia at upper left. A fifteenth-century editor, Nicolaus Germanus, drew the map using a method of projection Ptolemy explained elsewhere in the *Geographia.* The map of northeastern Europe from the 1482 Florentine edition is the eighth of 10 Ptolemaic maps showing various parts of Europe. This edition incorporates all of Ptolemy's traditional 27 maps, but replaces his ancient text with a contemporary geographical survey by Francesco Berlinghieri, written entirely in verse.

References: BERGGREN AND JONES; DILKE; SKELTON 1963 AND 1966B.

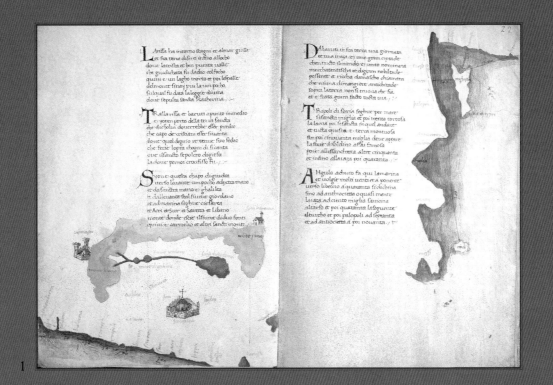

1

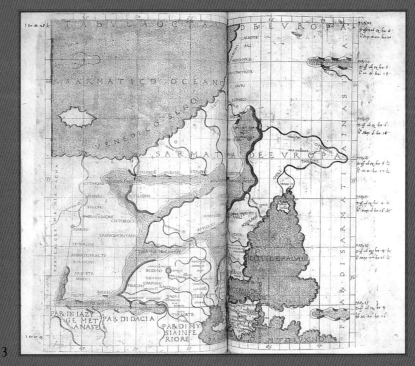

3

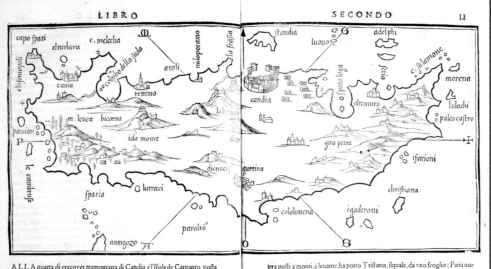

A L L A quarta di grecover tramontana di Candia é l'Isola de Carpanto posta che al presente Scarpanto é nominata, & da quella per migia cinquanta se dilóga & ésto nome, dalla quátita di frutti che qui naseono ageuolmente hauer cósegui/ to potrebbe. Quesla Isola e molto alta, & il nome di Carpatio al mare doue ella siede, gli dette, nella quale Palane de Titan figliuolo, hebbe sua habitatione, dal cui nome Palane ne fu anchora detta, & anchora quiui la Dea Pasa, fu nutrita . Et per lo adietro hebbe sei castella, de quali rimasti vi sono al presente, & so/

pra posti á monti, á leuante, ha porto Tristano, ilquale, da vno scoglio (Faria no/ minato) é fatto, & á ponente , Porto Grato tiene, doue per il tempo passato furo/ no, due castella, & presso al móte Gomello altresi due, l'vno Corezi detto. Questa Isola circoisse miglia settanta, & da garbino ha vna Isola Caso, nominata, & alchu/ n'altri scogli che Cani se appellano gli quali, háno, de circoito miglio vno , & so/ no del quarto clima nel principio, & al nono parallelo, & il suo piu longo giorno e di hore quattordeci & vno quarto.

K iii

4

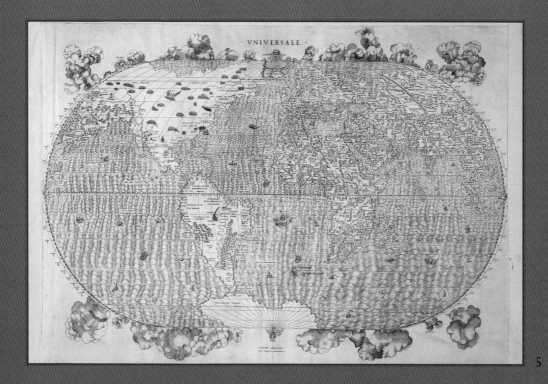

5

4   BENEDETTO BORDONE, *Isolario di Benedetto Bordone nel qual si ragiona di tutte l'isole del mondo* (VENICE: NICOLO D'ARISTOTILE, 1534). [AYER *7 B65 1534]

Extremely popular in the fifteenth and sixteenth centuries, the *isolario,* or "island book," has virtually ceased to exist, except perhaps in the form of the sailor's "cruising guide." Bordone's was one of the most widely printed examples of the genre. The Newberry possesses several editions, including the first from 1528. The *isolario* raises rich psychological questions. Did it represent an attempt to recapture the old, reassuring "world island" of the then-discredited T-in-O maps? Or was the island just more easily grasped, comfortable, and safe—in W. H. Auden's words, "like a lake turned inside-out"?

References: ARMSTRONG; SKELTON 1966A.

5   GIACOMO GASTALDI, *Universale* (VENICE: GASTALDI, 1546). [NOVACCO 4F 4]

Columbus's Transatlantic encounter and its aftermath stimulated the market for maps. And while Atlantic Europe's rise to dominance diminished Italy's position in European economics and politics, Italians remained at the center of European learning, fine arts, and publishing. The trade in maps as consumer items in their own right first flourished among engravers and printsellers based in Rome and Venice. Among these, the Venetian Giacomo Gastaldi earned the most enduring reputation as a skilled and conscientious cartographer. His striking world map of 1546 was the model for many world maps produced in Europe in the 1550s and 1560s. Gastaldi's map asserts a land connection between Asia and North America, a notion rejected by most cartographers by 1570 but not definitively refuted until the early eighteenth century.

References: WOODWARD 1996; SHIRLEY, NO. 85.

6   Battista Agnese, [Portolan atlas] (Venice, ca. 1550). [Ayer MS Map 12]

See color plate on page 84.

More than seventy manuscript atlases are attributed to Agnese, whose work flourished from 1536 to 1564. All are about the same size, all are exquisitely drawn and illuminated, and many include the same suite of maps. Here was a specialized craftsman (or workshop) catering to a market for handmade portolan-style maps in a period dominated by the printing press. His map of South America seems to draw on several sources, but owes a good deal to Gastaldi's *Universale* of 1546 (item 5). Though displaced too far inland, the Peruvian cities of Cajamarca, Huánuco, and Cuzco are nevertheless shown in more or less their proper relationship to one another.

References: Brown 1952; Wagner 1931 and 1947.

7   Giacomo Gastaldi, *Il disegno della geografia moderna de tutta la parte dell'Africa* (Venice: Fabius Licinius, 1564). [Novacco 6F 39]

In the 1550s, Giovanni Battista Ramusio published several previously unavailable and important travelers' accounts of the African continent. His associate, Giacomo Gastaldi, captured those accounts in cartographic form. A Venetian cartographer and engineer, Gastaldi could already boast a long and distinguished career when this map was published just two years before his death. Consisting of eight sheets (the Newberry's copy lacks the lower left corner), Gastaldi's was the largest map of the continent to date; other mapmakers frequently copied it.

References: Karrow 1993, no. 30/93.8; Norwich.

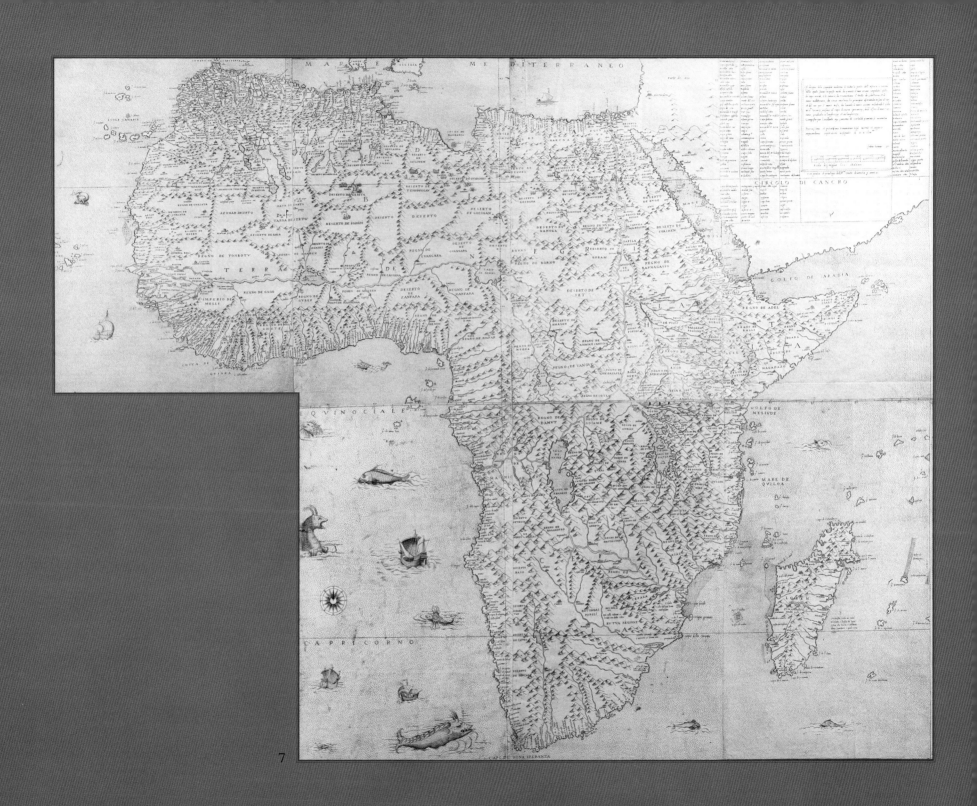

8

9

8 ABRAHAM ORTELIUS, "ASIAE NOVA DESCRIPTIO," IN HIS *Theatrum orbis terrarum* (ANTWERP: AEGIDIUS COPPENS DIESTH, 1570). [AYER *+135 O7 1570]

Ortelius's "Theater of the World," the first comprehensive modern world atlas, sparked a creative period in atlas production led by cartographers based in the Low Countries. The *Theatrum* included a catalogue of authors and sources unique among early atlases, explaining, for instance, that the map of Asia was based in part on the work of Giacomo Gastaldi. Ortelius's labors met with great commercial success: more than thirty editions of the *Theatrum* were published in seven languages between 1570 and 1612, along with an equivalent number of pocket-sized "epitomes." The Newberry holds 34 different versions of this landmark work.

References: AKERMAN 1995; BROECKE 1996, NO. 6; BROECKE ET AL. 1998.

9 GERARD MERCATOR, "SEPTENTRIONALIUM TERRARUM DESCRIPTIO," IN HIS *Atlas; sive, cosmographicae meditationes de fabrica mundi et fabricati figura* (DUISBURG: ALBERTUS BUSSIUS, 1595). [AYER *135 M5 1595]

Published in parts beginning in 1585, Mercator's *Atlas* gave its name to the genre. The third and final part, shown here, appeared in 1595, a few months after his death. As the most "scientific" cartographer of the sixteenth century, Mercator committed himself to presenting an accurate picture of the entire world. Because some regions of the globe had never been visited—much less mapped—by Europeans, he was forced to rely on whatever evidence existed, however flimsy. In the case of the area around the North Pole, this consisted of a textual account (now lost) by a fourteenth-century traveler who claimed to have been there. While we might dismiss Mercator's sources as utterly unsubstantiated hearsay, his contemporaries applauded his meticulous and judicious mapping of the best available "data."

References: KARROW 1993, PP. 376–406; OSLEY; WATELET, PP. 15–29.

10 Kaspar Vopel, *Europa. Das erste bewohnhafftige und fürnembste Dritte theil des Erdtreichs* (Cologne: Wilhelm Lützenkirchen, 1597). [Novacco 2F 21]

See color plate on page 85.

Two wall maps of Europe appeared in the middle of the six-teenth century: Gerard Mercator's 15-sheet copper engraving of 1554 (known in only one copy) and Kaspar Vopel's 12-sheet woodcut of 1555. No first edition copy of Vopel's map survives, and the Newberry's is the only known copy of the second edition, printed from the original woodblocks in 1597. Fully assembled, multisheet wall maps provided an excellent means of grasping the geography of a continent, but their extreme vulnerability to sunlight and general wear and tear meant that few would last in their original form. Most known examples of early wall maps survive, like this copy, only in their constituent sheets.

Reference: Heijden, pp. 58–61; Karrow 1993, pp. 386, 566–67.

11 Nicolas Sanson, "Amerique Septentrionale" (Paris, 1650), from his *Cartes générales de toutes les parties du Monde* (Paris: Pierre Mariette, 1658–[59]). [Case *+oG1015.S35 1658]

A key figure in France's emergence as a leading center of mapmaking in the middle of the seventeenth century, Nicolas Sanson served as geographer to Louis XIV. Sanson was a *géographe du cabinet*—an office-bound geographer who assessed and compiled the geographic information gathered in the field by others. The maps in his world atlas conformed to schematic divisions and subdivisions he laid out in a set of geographical tables for students in 1644. Here, on his map of North America, he carefully distinguished between major colonial territories. Sanson grouped New England and New Holland with New France as part of what Sanson labeled "Le Canada ou Nouvelle France."

References: Pastoureau 1984, pp. 400–17, and 1988; Pedley.

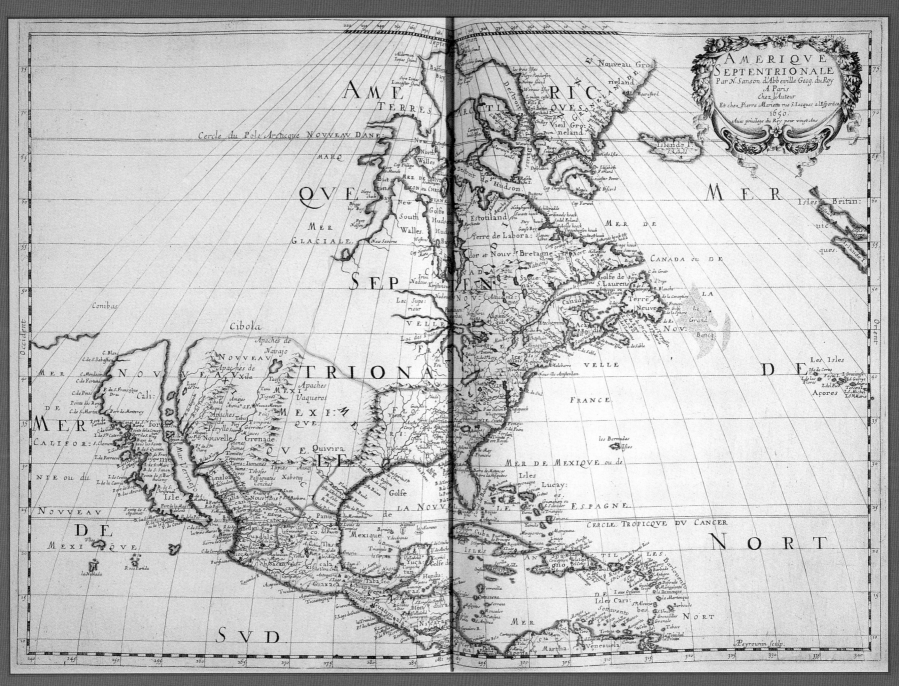

AMERIQVE SEPTENTRIONALE
Par N. Sanson d'Abbeville Geog. du Roy
A Paris
Chez l'Auteur
Et chez Pierre Mariette rue S. Iacques a l'Esperáce
1650.
Auec priuilege du Roy pour vingt Ans

11

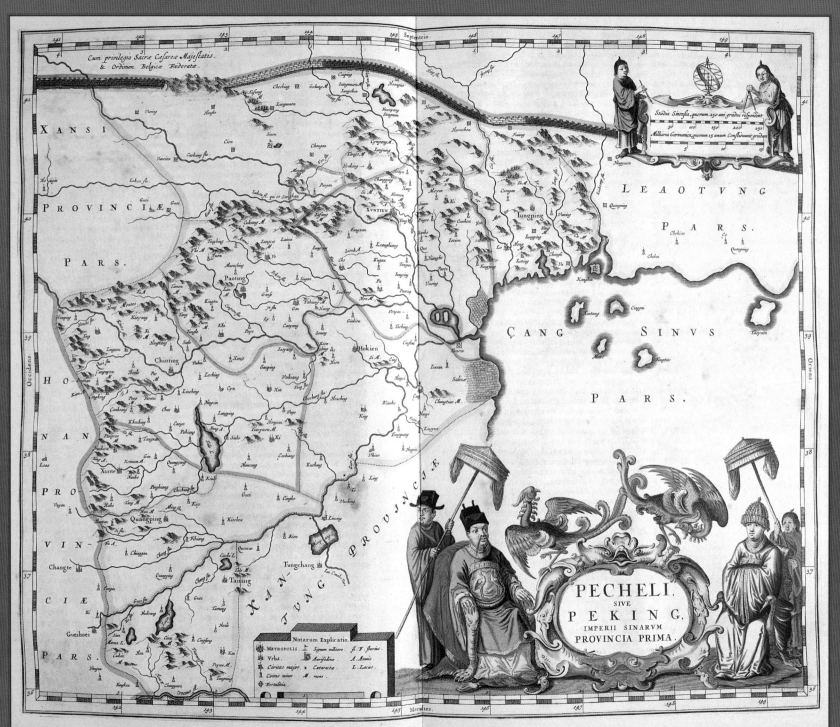

PECHELI.
SIVE
PEKING.
IMPERII SINARVM
PROVINCIA PRIMA.

12 MARTINUS MARTINI, "PECHELI SIVE PEKING," IN JOAN BLAEU, *Le grand atlas*, 12 VOLS. (AMSTERDAM: BLAEU, 1663), VOL. 11. [AYER *+135 B63 1663, VOL. 11]

Amsterdam publishers dominated the European atlas trade for most of the seventeenth century. Competition among them was acute, particularly between the publishing houses of Hondius-Jansson and the Blaeu family. Creating ever-larger atlases in their drive to outdo one another, they shamelessly copied each other's work and reused old map plates for decades. But their competition also brought to light many obscure or previously unpublished cartographic works. The atlas of China that Joan Blaeu began incorporating into his multivolume world atlases in 1655 is one such case. The maps were drawn by the Jesuit Martinus Martini, who seems, in turn, to have copied a sixteenth-century Chinese source. Here we see the district of Beijing with a portion of the Great Wall from the grand 12-volume atlas Blaeu issued in 1663.

References: KOEMAN 1967–85, 1: 68–70, 199–200, AND 1970; SZCZENIAK.

13 FREDERIK DE WIT AND GIUSEPPE LONGHI, *Nova totius terrarum orbis tabula* (BOLOGNA: GIUSEPPE LONGHI, CA. 1680). [NOVACCO MAP9F 1]

See color plate on page 86.

Wall maps became a popular form of domestic decoration in the sixteenth and seventeenth centuries and were frequently represented in contemporary paintings of interiors. Like its Dutch models, this Italian example (copied from a map published around 1660 by the Dutch cartographer Frederik de Wit) fills marginal spaces with geographical illustrations. Here, human figures representing Europe, Asia, America, and Africa stand in the foreground of scenes depicting the flora, fauna, and architecture thought to be typical of each continent. The map also perpetuates the mistake made by many Dutch cartographers, who depicted California as an island.

Reference: SHIRLEY, NO. 471.

14 *Géorama* (Paris: Le Fuel, between 1821 and 1832?). [Case oGV1485.G4]

At the heart of this early nineteenth-century set of 310 cards lies the theory, embraced by generations of European and American educators, that students could best understand world geography through rote memorization. Each country's vital statistics are laid out in five cards, prototypes of our modern flash cards. The first four cards list major geographical features, as well as facts about local culture and language, form of government, religion, and the economy. A fifth card reproduces an exquisite small map, with just enough room to delineate the outline of the country and a handful of its cities, mountains, and rivers. *Géorama* came to the Newberry as part of a large gift from Andrew McNally III that includes many unique examples of school geographies, games, and learning aids.

15 Great Britain, Hydrographic Office, *Chart of the South Polar Sea* (London: Hydrographic Office, 1841). [Map3C 8]

John Edward Davis carried this map south in 1841 while serving as second master to the *Terror,* one of two ships belonging to Sir James Clark Ross. Ross entrusted Davis with the job of charting the expedition. The manuscript annotations seen here belong to Davis and show the precise course of both ships along with their daily locations. The inset at upper right, entirely in manuscript, shows a portion of the Ross Sea coast, discovered and named on this expedition. Sir Clements Markham, an early historian of polar exploration, wrote that "old Davis . . . was a good artist, a good writer, a good surveyor, and a right good fellow." The seventh continent, Antarctica was the last to be understood by cartographers.

Reference: Ross, pp. 33–34.

14

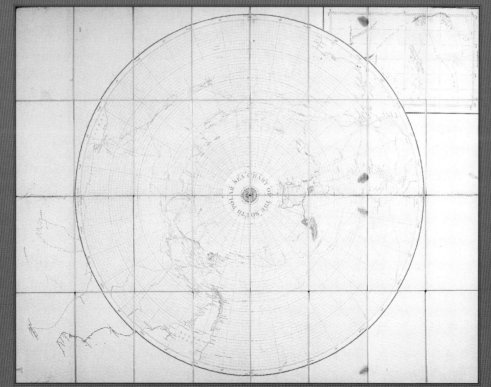

15

16

17

16 SIDNEY E. MORSE AND SAMUEL BREESE, "MAINE," IN THEIR *Cerographic Atlas of the United States* (NEW YORK: SIDNEY E. MORSE, 1842). [+G1083.59]

Distributed as a free supplement to new long-term subscribers of the *New York Observer,* this atlas quietly marked a revolution in American map printing. Its author, Sidney E. Morse (son of geographer Jedediah Morse and brother of the telegraph inventor), was one of several early developers of cerography, or wax-engraving, a process that made printing detailed but inexpensive maps possible. The technique never caught on in Europe, where cerographic maps were thought to be plain and inferior in workmanship. Decades later, Chicago publishers Rand McNally and George F. Cram relied on the process to produce cheap but functional railroad maps, as well as school and reference atlases for an emerging mass market.

References: PETERS; WOODWARD 1977.

17 SAMUEL AUGUSTUS MITCHELL, *Mitchell's Dissected Map [of the] U.S.* (PHILADELPHIA: COWPERTHWAIT, DESILVER & BUTLER, 1854). [MAP4C oG3701.A9 1854 .M5]

The first jigsaw puzzles date from the 1760s and were made from maps. Assembling a complete map from a pile of small, distinct pieces has been a popular and challenging pastime ever since. S. A. Mitchell was the premier American cartographer of his day, and he intended his series of "dissected" maps "to make the young people familiar with the locality of the different parts of their own country." Mitchell's series included four regional maps of the United States but, curiously, no world map. Would-be cheaters are warned: "the Key Map should not be seen or referred to, until a patient effort has been made to put the map together without it."

Reference: SHEFRIN.

18 DENNIS TOWNSEND, *Townsend's Patent Folding Globe* (BOSTON: GEORGE M. SMITH & CO., 1869). [MAP4C 0G3170 1869 .T6 COPY 1 OF 2]

Vermont is the birthplace of American globe making and of Dennis Townsend, the inventor of this clever and effective folding globe. Forced to leave Dartmouth after two years owing to "pecuniary embarrassments," Townsend joined the gold rush to California, where he became a teacher and superintendent of schools. The University of Vermont gave him an honorary A.M. in 1869. While every cartographer since Ptolemy has recognized the superiority of a globe for studying and teaching world geography, the size, awkwardness, and expense of globes have necessarily restricted their use. For at least the last two hundred years, a variety of collapsible globes, made of paper, cloth, gutta-percha, and plastic, has been fashioned to help solve this problem.

Reference: LANMAN 1983.

19 R. BUCKMINSTER FULLER, *Dymaxion Airocean World* (RALEIGH: SCHOOL OF DESIGN, NORTH CAROLINA STATE COLLEGE, 1954). [FITZGERALD MAP6F 0G3201.B72 1954 F8]

The twentieth century's most influential futurist and global thinker, "Bucky" Fuller coined the term "spaceship earth," designed globelike buildings, and doggedly tried to disabuse people of the persistent illusion that they lived on a flat planet. His answer to the "folding globe" challenge was the Dymaxion projection, in which the equilateral triangle provides both mathematical precision and structural strength. As a flat map, it stresses the connectedness of the continents, forming, as in the T-in-O maps and *isolario,* a "world island." Fuller saw his map, first published in *Life* magazine in 1943, as a tool for global problem solving. He patented it in 1946.

Reference: SNYDER, PP. 269–70.

The word Dymaxion and the Fuller Projection Dymaxion™ Map design are trademarks of the Buckminster Fuller Institute, Sebastopol, California. *Raleigh Edition* ©1943, 1944, 1953, 1954, 1967, 1971, 1980, 1982. All rights reserved.

18

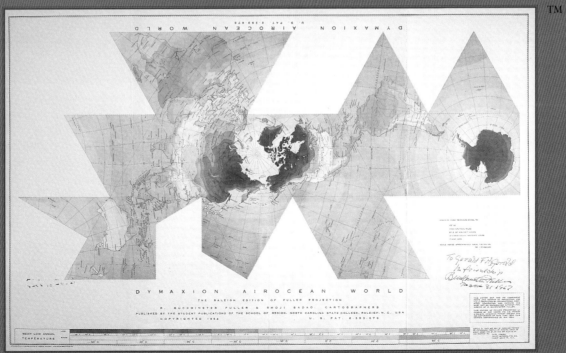

19

INVENTING THE NATION

20 SEBASTIÃO LOPES, [CHARTS OF THE BAY OF BENGAL AND SOUTHEAST ASIA], IN HIS [PORTOLAN ATLAS] (LISBON, CA. 1565). [AYER MS MAP 26]

See color plate on page 87.

The 24 richly illustrated charts in this magnificent manuscript atlas convey a powerful sense of the global reach exerted by the Portuguese maritime empire at its height. Read in order, they plot a voyage around the world beginning in the Mediterranean Sea, then passing southward through the Atlantic Ocean, around the Cape of Good Hope, into the Indian Ocean, across the Pacific to the Americas, and home again. Together, the two charts reveal the main target of Portuguese mariners: the Bay of Bengal and the spice-producing islands of the East Indian archipelago.

References: CORTESÃO AND TEIXEIRA DA MOTA, VOL. 4, PP. 9–14, PLS. 389–96; SMITH, NO. 26.

21 PHILIPP APIAN, [INDEX MAP] FROM HIS *Bairischen Landtafeln XXIIII* (INGOLSTADT: APIAN, 1568). [AYER *7 A71 1568]

A professor of mathematics at the University of Tübingen, Philipp Apian traveled through Bavaria measuring the distances between towns with an odometer and using a compass to record bearings. The result, in the form of a 24-sheet atlas, was a very early instance of instrumental surveying used to map an entire country. If assembled, the map would measure about five feet square. The sheet shown here is the earliest known example of an index map; it indicates the content and arrangement of each sheet in the atlas. The original woodblocks from which the map was printed still exist in the Bavarian National Museum.

References: KARROW 1993, PP. 64–70; WOLFF.

22 CHRISTOPHER SAXTON, "NORTHAMTON, BEDFORDIAE, CANTABRIGIAE, HUNTINDONIAE, ET RUTLANDIAE COMITATUUM . . . 1576," IN HIS *Atlas of England and Wales* ([LONDON], 1579). [CASE *+G 1045.78]

See color plate on page 88.

One of the most ambitious cartographic projects of the sixteenth century, Saxton's painstaking county-by-county mapping of England became a prototype for modern national atlases and surveys. Queen Elizabeth I's principal minister, Lord Burghley, instigated the project; a relatively minor crown official named Thomas Seckford then privately financed it. Saxton spent nine years compiling maps both from his own surveys and from existing maps and local records. Aside from towns and local boundaries, his maps record items of particular interest to the queen and her administrators, such as bridges and royal forests (indicated by small round enclosures). The queen's arms appear at upper left, just above the title cartouche, while Seckford's are at lower left. The cartographer's name is barely visible on a simple line of text just above the scale.

References: EVANS AND LAWRENCE; TYACKE AND HUDDY.

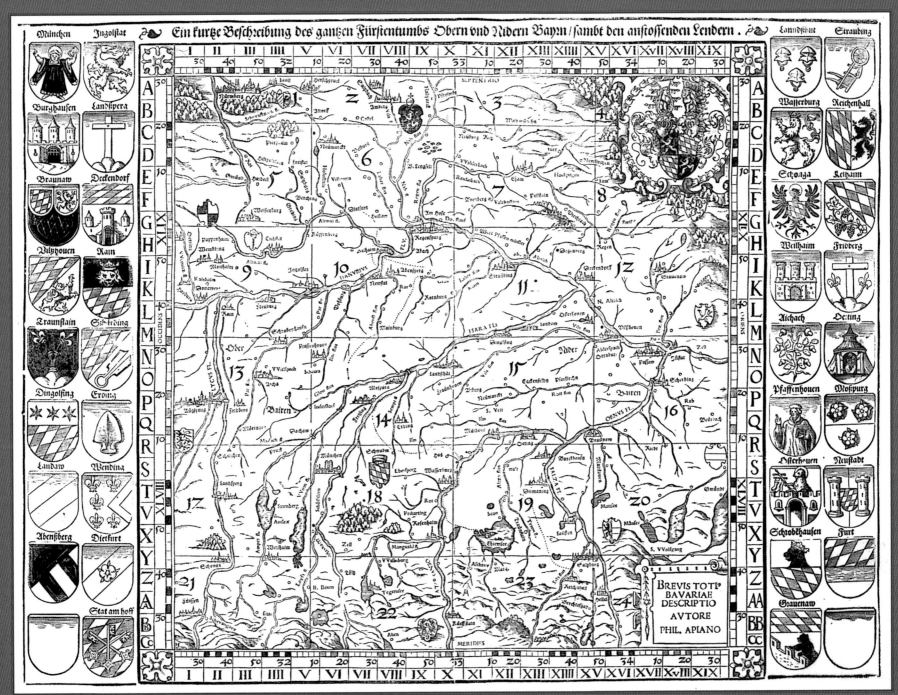

Ein kurtze Beschreibung des gantzen Fürstentumbs Obern vnd Nidern Bayrn / sambt den anstossenden Lendern.

München Ingolstat Landßhut Straubing
Burghausen Landßpera Wasserburg Reichenhall
Braunaw Deckendorf Schonga Kelhaim
Vilßhouen Rain Weilhaim Fridberg
Traunstain Scherding Aichach Oeting
Dingolfing Erding Pfaffenhouen Moßpurg
Landaw Wending Osterhouen Neustade
Abenßberg Dietfurt Schrobenhausen Furt
Stat am hoff Grauenaw

BREVIS TOTI
BAVARIAE
DESCRIPTIO
AVTORE
PHIL. APIANO

21

23

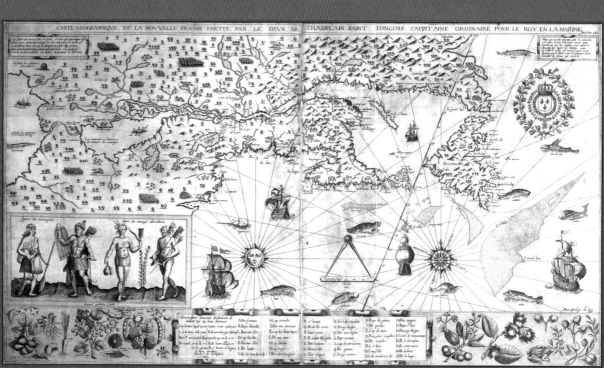

24

23 GERARD MERCATOR AND ABRAHAM ORTELIUS, "ERYN. HIBERNIAE BRITANNICAE INSULAE, NOVA DESCRIPTIO," IN ORTELIUS, *Theatro de la tierra universal* (ANTWERP: PLANTIN, 1588). [AYER *+135 O7 1588]

After publishing the first edition of his *Theatrum orbis terrarum* in Latin (item 8), Abraham Ortelius and his publisher broadened its audience by issuing the atlas in several vernacular languages, including this Spanish edition of 1588. In addition, Ortelius regularly updated his maps with new ones sent from correspondents across Europe. As the atlas expanded, readers from almost every part of the continent could find in it regional maps that appealed to their local pride. The 1570 edition, for instance, included only a single map of the British Isles, but Ortelius added separate maps of England, Wales, Scotland, and Ireland to all editions published after 1572. He copied this map of Ireland from a 1564 Gerard Mercator map of the British Isles.

References: ANDREWS 1997; BROECKE 1996, NO. 22; KARROW 1993, PP. 1–31; WATELET.

24 SAMUEL DE CHAMPLAIN, *Carte geographique de la Nouvelle Franse* (PARIS, 1612). [AYER *MAP4F OG3300 1612 .C5]

Champlain could be said to have invented New France; his maps provided a basis for France's claims on the American continent. One of the foundations of North American cartography, this map depicts areas Champlain visited or received firsthand reports of during his long residence in New France. In the last 30 years of his life, Champlain made no fewer than 23 Atlantic crossings, and his travels took him as far south as Martha's Vineyard and as far west as Montreal. For information beyond Montreal, his map relies on the observations of his lieutenant, Etienne Brûlé, who may have journeyed as far west as Lake Superior.

References: HEIDENREICH; MORISON.

25 JOHN FARRAR, "A MAPP OF VIRGINIA," IN EDWARD WILLIAMS, *Virginia; more especially, the south part thereof* (LONDON: PRINTED BY T. H. FOR J. STEPHENSON, 1650). [AYER *150.5 V7 W7 1650]

Seeking to attract prospective English investors, promoters of the crown colony of Virginia (so named in 1624) avidly publicized the area's natural bounty. Edward Williams's account spares few superlatives: "Nature regards this Ornament of the new world with a more indulgent eye than she hath cast upon many other Countrys." He goes on to compare Virginia to Persia, predict the discovery of rich silver mines, and opine that "the South west Passage may easily be found out by a constant intelligence and information of the Natives: from whence a trade and commerce may be driven with China and Cathaya." The last of these advantages to Virginia's "nationhood" is a central feature of John Farrar's map, which asserts that Drake's "New Albion," on the shores of the Pacific, can be reached "in ten dayes march with 50 foote and 30 horsmen from the head of Jeames River. . . ."

Reference: VIRGINIA, NO. 6.

26 ALEXIS HUBERT JAILLOT, *Galliae Regnum. Le Royaume de France* (AMSTERDAM: JAN COVENS AND CORNEILLE MORTIER, CA. 1725). [SACK MAP8F 0G5830 1725 .J3]
See color plate on page 89.

From about 1720 until his death in 1751, Baron Johann Gabriel Sack, a Swedish nobleman with familial ties to diplomatic circles, assembled a magnificent collection of roughly six hundred maps that reflected his interest in European political affairs. The Sack Collection includes two nearly identical wall maps of France dating from the 1720s, one published in Paris by Jaillot and this one published in Amsterdam by Covens and Mortier. Both maps show the territory Louis XIV annexed to France's northern, western, and southern frontiers, and both maps include ample portions of German, Dutch, and Italian territory—stages for the Sun King's military adventures. The king is here depicted in conversation with Mars, the god of war, who gazes at a map of the British Isles, France's longtime nemesis. The Dutch publishers added the colorful title vignette at upper left, perhaps as a subtle critique of Louis's aggressiveness.

Reference: BOSSE 1982A.

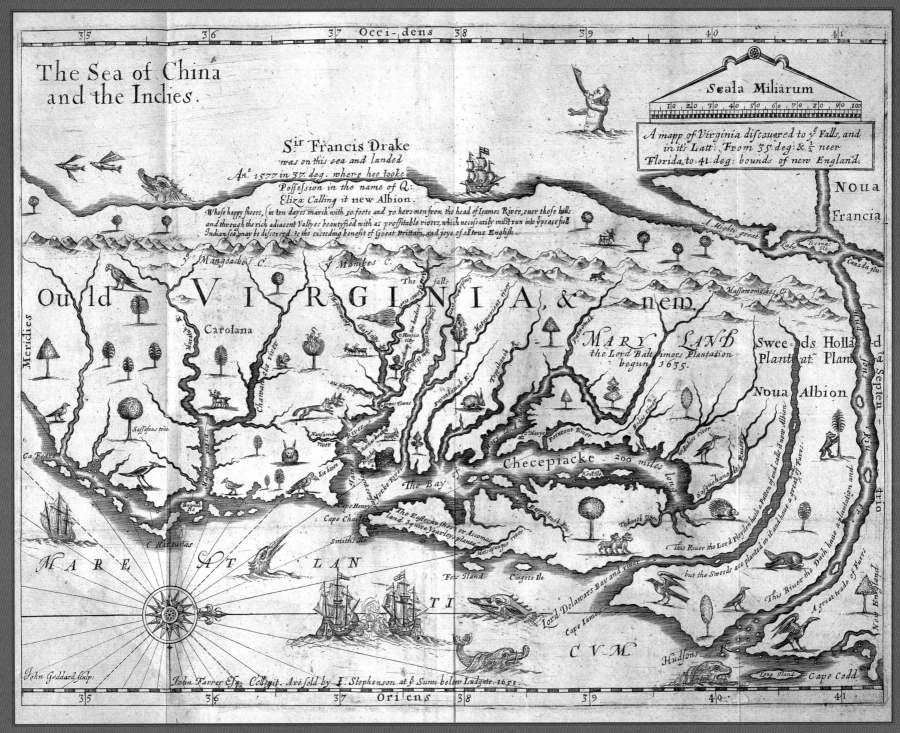

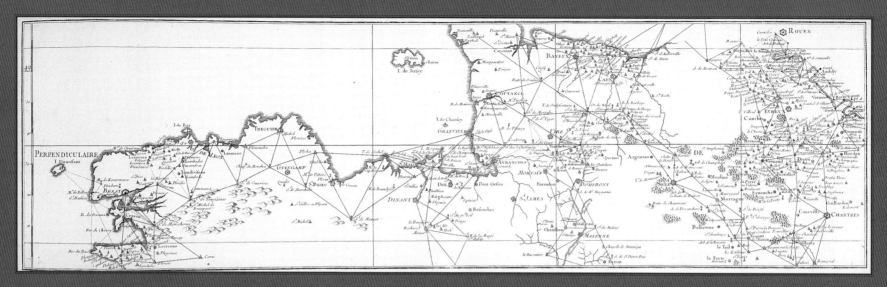

27 *(detail)*

28

27 GIOVANNI MARIA CASSINI, *Carte qui comprend touts les lieux de la France qui ont été déterminés par les opérations géometriques* (PARIS, 1745). [CASE *7Q 89]

With the completion of the great Cassini map in 1789 (more than a hundred years after it was begun), France became the first nation to be represented by a uniform, detailed, topographical map based on an instrumental survey. To fill in the map's 180 sheets of meticulous topography required a geometrical framework, built from 800 interconnecting triangles whose angles had been accurately measured from atop church towers and other prominent landmarks. This is one sheet of the map recording that completed triangulation network. Upon seeing a preliminary version of the survey, in which the outlines of his kingdom appeared to have shrunk, Louis XIV is said to have told the cartographers: "Your work has cost me a large part of my state!"

References: BROWN 1949, PP. 241–55; KONVITZ; PELLETIER, P. 11.

28 GEORG HAAS, *Carte des partages de la Pologne en 1772, 1793 et 1795* (BASEL: G. HAAS AND J. DECKER, CA.1800). [MAP4F oG5701.A1 1777 .H3]

With the expansion of Europe's educated classes during the eighteenth century, public discussion of politics, science, and history flourished, fueling demand for maps. To satisfy an expanding readership, publishers began experimenting with ways to prepare and print maps more rapidly and cheaply than by means of copper engravings and woodcuts. One early experiment used "alphabets" of movable metal type—similar to that used for print text—to represent graphic elements such as mountains, towns, roads, rivers, and boundaries. This image, depicting the partition of the once-powerful Kingdom of Poland, is included in the Newberry's collection of typographic maps published by Wilhelm and Georg Haas. The diplomatic maneuver represented here preserved a fragile peace between Austria, Prussia, and Russia, but also sowed the seeds of Polish nationalism.

References: AKERMAN 1991–92; HARRIS.

29 HENRY S. TANNER, "GEORGIA AND ALABAMA," IN HIS *A New American Atlas* (PHILADELPHIA: H. S. TANNER, 1823). [CASE +oG1200.T3 1823]
See color plate on page 89.

Henry S. Tanner intended his ambitious *New American Atlas* both to rival European productions and to aid Americans' understanding of their national territory. His map of Georgia and Alabama shows the Southeast during a period of warfare, political uncertainty, and rapidly expanding non-native settlement. A bold yellow line delineates the boundary between the two states, but also bisects the territory of the Cherokees and Creeks, who remain outside of the political geography of organized counties that are pressing in upon them. In the 1830s, both Indian nations were forced to accept treaties ceding almost all of their remaining lands and removing most of their population via the Trail of Tears to what is now Oklahoma.

Reference: RISTOW 1997.

30 WILLIAM CHAPIN, *Chapin's Ornamental Map of the United States* (NEW YORK: CHAPIN & TAYLOR, 1845). [MAP8F oG3700 1845 .C5].

Chapin's wall map depicts the nation on the eve of war with Mexico. With the established county governments compactly filling the eastern states and seeming to press across the Mississippi, the map can be read as a graphic demonstration of "Manifest Destiny." The presidential portraits supply a pedigree, a historical justification for a country barely sixty years old. Funding for the new national map was raised in the time-honored way: by subscription. The Newberry is fortunate in having the original subscription book for the first edition of the map, which bears the names and addresses of more than eight hundred subscribers in New York, Philadelphia, and Baltimore. Taylor, the map's copublisher, combed the streets of the business districts gathering signatures of such captains of commerce and industry as banker August Belmont. The subscription price of $9 would be about $150 today.

31 JOHN DISTURNELL, *Mapa de los Estados Unidos de Méjico* (NEW YORK: J. DISTURNELL, 1847). [GRAFF 1092] *See color plate on page 90.*

A New York publisher of reference books, handbooks, guidebooks, and maps, John Disturnell had the good fortune to issue his map of Mexico in 1846, just as the Mexican-American War began. Though Disturnell based his map on one published by Henry S. Tanner in 1825, he updated political boundaries such as that of the independent Republic of Texas and added inset maps of places connected to wartime events. By the war's end, Disturnell had to publish several more editions to meet demand. This became the map of record attached to official copies of the 1848 Treaty of Guadalupe-Hidalgo, by which Mexico lost roughly one-half of its prewar territory to the United States.

References: MARTIN; WHEAT, VOL. 3, PP. 35–37.

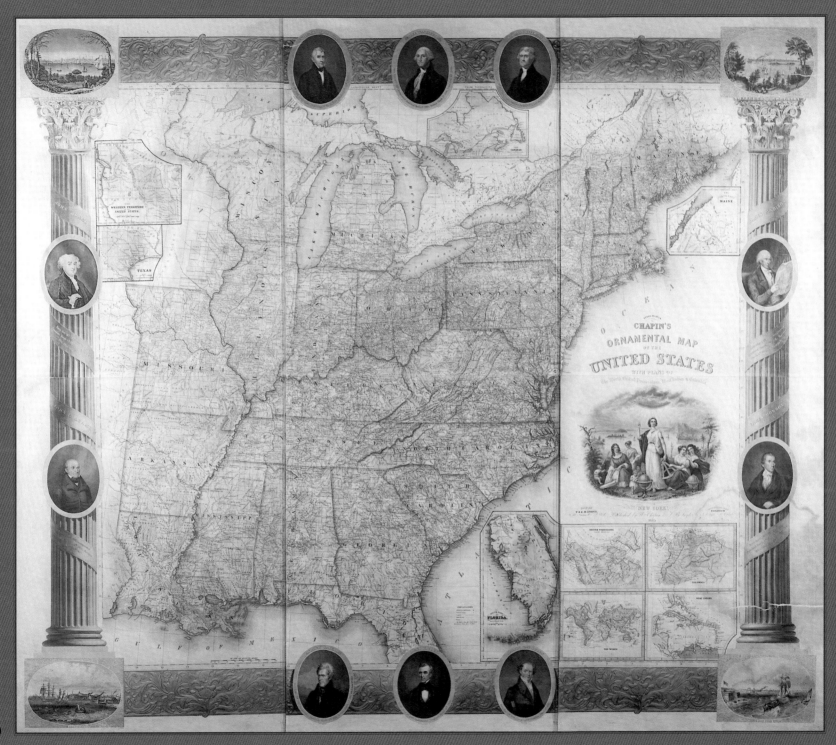

30

32   F. W. von Egloffstein, "Rio Colorado of the West," from Lt. Joseph C. Ives, *Report upon the Colorado River of the West* (Washington: Govt. Printing Office, 1861). [Ayer map6F oG4302.C6 1858 .I91]

Having reshaped the United States after the Mexican War with the addition of a million square miles, the federal government set out to investigate one of the greatest natural resources of the new southwestern territories, the Colorado River. Starting in the Gulf of California, the expedition worked its way upstream to the head of navigation, near present-day Hoover Dam. Topographer Egloffstein's party then went overland, finally reaching the reported "Big Cañon." While impressed by its size, Joseph Ives concluded that the Grand Canyon "is, of course, altogether valueless. It can be approached only from the south, and after entering it there is nothing to do but leave. Ours has been the first, and will doubtless be the last, party of whites to visit this profitless locality."

Reference: Wheat, vol. 4, pp 95–101.

33   William Henry Holmes, "The Grand Cañon at the Foot of the Toroweap — Looking East," in the atlas accompanying Clarence E. Dutton's *Tertiary History of the Grand Cañon District* (Washington: Govt. Printing Office, 1882). [Ayer 109.9 G4 D9 1882 Atlas]
See color plate on page 90.

One of the greatest topographers of the nineteenth century, William H. Holmes cut his teeth on Ferdinand Hayden's western geological surveys of the 1870s. While curator of anthropology at Chicago's Field Museum in the 1890s, he made splendid illustrations of ancient sites in the Yucatan. Holmes's

Grand Canyon views, drawn during his time with Dutton on the new U.S. Geographical Survey, capture the grandeur that maps alone could not express. Twenty years after this view was published, Fred Harvey, pioneer of western tourism, was bringing sightseers to the South Rim, despite Ives's gloomy prediction (item 32).

References: Fernlund; Goetzman, pp. 512–13.

34   Count Pál Teleki, "Ethnographical Map of Hungary," from his *The Evolution of Hungary and its Place in European History* (New York: Macmillan, 1923). [Map6F oG6501.E1 1923 .T4]
See color plate on page 91.

Since the nineteenth century, geographers have compiled ethnographic maps in an attempt to represent the distribution of different cultural or ethnic groups. The collapse of the multinational empire of Austria-Hungary after World War I posed a particular problem for peacemakers determined to redraw the political map of Eastern Europe based on the slippery concept of ethnic or national identity. The Hungarian delegates to the peace conference, Count Teleki among them, argued against this approach, noting that most ethnographical maps were prone to oversimplification. A geographer at the University of Budapest and a future prime minister of Hungary, Teleki attempted to reflect local ethnic diversity, particularly in cities, in his own maps. This colorful example was published along with a general history of Hungary.

Reference: Richardson.

CONTESTING PLACES

35 *Stadt Pavia* (BASEL?, 1525?). [CASE WING +oG6714. P45 1525 .S8]

An early example of a "news map" intended to illustrate an event shortly after its occurrence, the map shown here depicts the battle of Pavia, fought on 24 February 1525. Pitting a French army of 28,000 men against a slightly smaller Habsburg force, the battle became a wholesale disaster for the French. After leading his troops into battle like a medieval king, Francis I was carried off to Madrid as a prisoner; France surrendered her claims to Italy; and the era of Spanish hegemony began. Swiss and German mercenaries fought alongside the French and may have carried this battle map back home to be printed, probably in Basel. The basic plan was made before the battle's start, with a few details (some of them incorrect) added as an afterthought. The map remained totally unknown to historians until 12 copies were discovered in a binding in the early 1990s. The sheets had been culled from the printer's scrap pile and were used to stiffen the boards of a book printed in 1529 and probably bound shortly thereafter.

References: GIONO; SINISTRI AND CASALI.

36 ALEXANDER DE GROOTE, *Neovallia dialogo . . . nel quale con nuova forma di fortificare piazze s'esclude il modo del fare fortezze alla regale, come quelle che sono di poco contrasto* (MUNICH: VEDOUA ANNA BERGHIN, 1617). [CASE FU 26.374]

Military engineers play a crucial role in the history of cartography. The profession's origins lie in the fifteenth century, when improved cannons demanded a science of ballistics and stronger defenses to withstand their assaults. Engineers had to master mathematics and geometry, arts that were also prerequisites for the emerging discipline of modern cartography. By the seventeenth century, when the manual shown here was published, the design of fortifications had reached the level of high art. One historian has said that de Groote's plates "are distinguished by their aggressive bellicose expressiveness," traits well illustrated in this lively, but hypothetical, battle.

Reference: POLLAK, PP. 54–55.

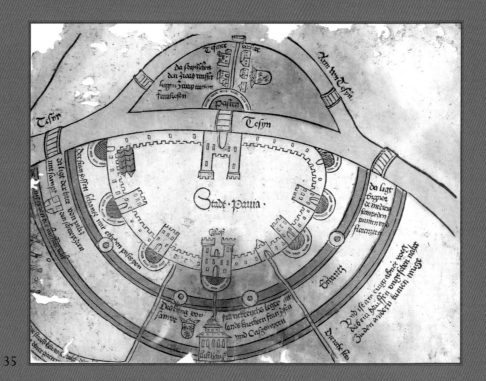

35

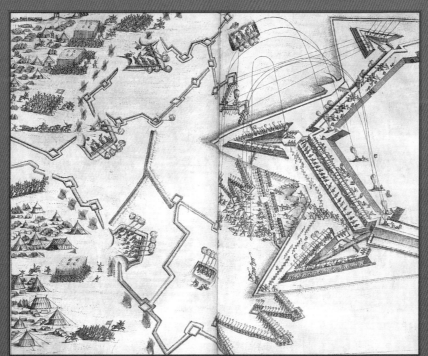

36

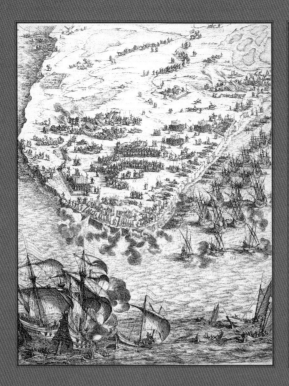

37 *(detail)*

38

37 JACQUES CALLOT, *La Siège de la Rochelle* (PARIS, 1629). [NOVACCO 4F 112]

Europe was beset by wars throughout much of the seventeenth century. These conflicts, besides providing employment to military engineers and cartographers, inspired fine artists to document the carnage. No printmaker ever captured the confusion, agony, and ecstasy of battle more vividly than the French etcher Jacques Callot. Callot produced more than fourteen hundred prints and two thousand drawings, many of them religious, but three large battle scenes form an important part of his oeuvre as well. Commissioned by Louis XIII, this print depicts the king's capture of La Rochelle—a notorious Protestant stronghold on the Atlantic coast of France—from the Huguenot army. Callot's fine technique, coupled with an improved method of preparing the ground, enabled him to produce an immense range of illumination, lending his landscapes exceptional depth and spaciousness. He was also known for a great richness of detail, and this illustration of the siege convinces the viewer that every soldier has been individually rendered.

References: BECHTEL; RUSSELL.

38 CLAES JANSZOON VISSCHER, *Nieuwe kaerte van Breda belegert . . . den 23 Iuly 1637* (AMSTERDAM: VISSCHER, 1637). [VISSCHER COLLECTION, NO. 28]

Siege warfare dominated the European conflict known as the Thirty Years' War (1618–48), as well as the war within the war between the Netherlands and Spain. When Spanish troops captured the southern fortress city of Breda in 1625, they posed a major threat to the survival of an independent Netherlands. This siege plan narrates the events leading up to the city's recapture in October 1637 by forces under the command of Dutch stadtholder Prince Frederick Henry. Visscher published siege plans during the war, several of which (as here) were accompanied by brief news accounts of the siege. The Newberry's Visscher Collection includes 24 of these early "news maps."

Reference: CAMPBELL 1968.

39 JOHN MITCHELL, *A Map of the British and French Dominions in North America* (LONDON, 1755). [AYER 133 M66 1755]

See color plate on page 91.

While the Mitchell map does not actually show a battle, it graphically illustrates the contest for the continent played out in the French and Indian War (also known as the Seven Years' War). The broad colored swaths sweeping west show a British determination to enforce the "sea to sea" charters of the southern colonies, while preventing further encroachments by "New France or Canada," here colored green. On the strength of treaties with the Iroquois Confederacy, the British confidently depicted an "international border" beginning at the bottom of Lake Michigan and heading southwest into Illinois Country. This is the third impression of the first edition, a version produced recently enough that the coloring of Acadia (New Brunswick) reflects its June 1755 capture by the British.

References: RISTOW 1972; SCHWARTZ, PP. 68–69.

40A & B *Bataille d'Hastembeck, 2e disposition* (HASTENBECK, GERMANY, 1757). [MAP4F 0G5701.55A. M5, NOS. 98–99]

A global conflict, the Seven Years' War saw action on four continents. In Germany, forces of several German states combined under the leadership of the English duke of Cumberland to face French troops led by the Count d'Estrées. The two armies met on a narrow strip of land situated between a marshy stream and the wooded heights above the town of Hastenbeck, near Hannover. The fighting lasted the better part of three days, and in the end, "the troops on neither side were beaten. Both commanders thought themselves defeated; but d'Estrées was the first to realize his mistake." While Callot's view of La Rochelle (item 37) telescopes three separate events into one picture, the anonymous French engineer at Hastenbeck made three maps to show various stages of one battle (two versions of the second are shown here). Even within these shorter periods, the complexity of troop movements posed a severe challenge to the cartographer. Dotted lines, explanatory legends, multiple maps, and overlays were all necessities in the military cartographer's kit. The cruder of these two manuscript maps was probably made shortly after the battle; the finished copy is clearly an office job.

Reference: SAVORY, PP. 25–46.

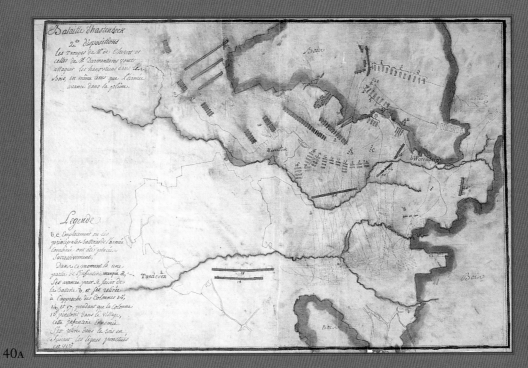

Bataille d'hastenbeck
2.ᵉ dispositions
Les Troupes de M.ᵉʳ de Chevet et
celles de M. Darmentieres vont
attaquer les hanoviens dans les
Bois, en même tems que l'armée
avance dans la plaine.

Legende
b.c. Emplacement ou les
principales batteries de l'armée
combinée ont été placées
Successivement;
Dans ce moment là une
partie de l'Infanterie marqué a,
s'est avancée pour le saisir de
la Batterie b. et s'est retirée
à l'approche des Colonnes 16;
14 et 17 pendant que la Colonne
18 penetroit dans le Village,
cette Infanterie ennemie
s'est retirée dans le Bois en
suivant les lignes ponctuées
en verd.

Tundeven

40A

Bataille d'hastembeck
2.ᵉ disposition.
Les Troupes de M.ʳ de Chevet et
celles de M. Darmentieres vont
attaquer les ennemis dans les
Bois, en même tems que l'armée
avance dans la plaine.

Legende
b.c. Emplacement ou les
principales batteries de l'armée
combinée ont été placées
Successivement.
Dans ce moment les deux
partie de l'Infanterie marqué a
s'est avancée pour le saisir de
la batterie b. et s'est retirée
à l'approche des Colonnes 15.
16 et 17 pendant que la colonne
16 penetront dans le Village,
cette Infanterie ennemie
s'est retirée dans le bois en
suivant les lignes ponctuées
en verd.

Tunderen

40B

A PLAN of
YORK TOWN AND GLOUCESTER,
IN THE PROVINCE OF VIRGINIA,
Shewing
the Works constructed for the Defence of those Posts
by the BRITISH ARMY,
under the Command of Lt. Genl. EARL CORNWALLIS,
together with
the Attacks and Operations of the American and French Forces,
Commanded by
Genl. WASHINGTON and COUNT ROCHAMBEAU,
to whom the said Posts were Surrendered
on the 19th October 1781.
from an actual SURVEY in the Possessions of
Jno. HILLS, late Lieut. in the 23 Regt. & Ast. Engr.

41

LLOYD'S MAP
of the
LOWER
MISSISSIPPI RIVER
FROM ST. LOUIS and the GULF of MEXICO,
COMPILED FROM GOVERNMENT SURVEYS IN THE TOPOGRAPHICAL BUREAU,
WASHINGTON, D.C.

42

41 WILLIAM FADEN, *A Plan of Yorktown and Gloucester* (LONDON: FADEN, 1785). [MAP4F oG3701.S3.29]

London publisher William Faden documented many of the Revolutionary War battles in printed maps, sometimes producing them within a month of the engagement. When Cornwallis surrendered his army to General Washington in October 1781, the British crown lost its tenuous hold on the rebellious American colonies. In that battle—depicted here by Faden—a powerful French force ensured Washington's victory: Admiral deGrasse secured the York River approaches while Rochambeau and Lafayette strengthened Washington's troops encircling Yorktown. During the peace settlement negotiated the following year, on a copy of the then-30-year-old Mitchell map (item 39), commissioners drew boundary lines with the new United States on one side and British possessions in Canada on the other.

References: HARLEY et al. 1978, PP. 79–110; NEBENZAHL 1974, PP. 185–91; NEBENZAHL 1975, NO. 195.

42 JAMES T. LLOYD, *Lloyd's Map of the Lower Mississippi River from St. Louis to the Gulf of Mexico, 1862* (NEW YORK: J. T. LLOYD, 1862). [MAP8F oG4042.M5 1862 .L55]

Though the title of this map does not explicitly refer to the American Civil War, it was issued just as the struggle for control of the river reached its critical phase. In April 1862, Union forces captured New Madrid, Missouri, and New Orleans (at opposite ends of the map, on sheets 1 and 5, respectively), and Memphis fell in June. After these, the only remaining Confederate river crossing that linked railroads on both banks was Vicksburg (at the top of sheet 4), which would defy Grant's efforts to capture it until July 1863. The plantations on the river's southern reaches were deliberately long and narrow to provide better access to the water, a legacy of practices established when the Mississippi Valley was a French colony. The map's five sheets could be joined to form one long, continuous map of the river. Then, installed on rollers in a special box, the map could be mounted on the wall of a riverboat's pilothouse. Lloyd acknowledges the help of pilots Bart and William Bowen, both of whom worked with Samuel Langhorne Clemens (later known as Mark Twain) on the Mississippi.

References: BOSSE 1982B.

43 JOHN B. BACHELDER, "GETTYSBURG BATTLEFIELD," FROM HIS *Gettysburg: What to See and How to See It* (BOSTON: BACHELDER, 1873). [MAP6F oG3824.G4.S5 1863 .B3]

When President Lincoln dedicated Gettysburg National Cemetery on 19 November 1863, he transformed a battlefield into a tourist shrine. Maps such as the one seen here, made in the wake of conflict, served the needs of pilgrims, survivors, and mourners drawn to battlefield sites to make their own personal peace with the past. This panoramic map of the battlefield was published at the Civil War's end, along with a guidebook for tourists prepared by Col. John Bachelder, then regarded as the nation's leading authority on the Battle of Gettysburg. Bachelder's re-creation of the events of July 1863 bears the endorsement of General Meade, the Union commander, as well as several of his lieutenants.

Reference: FRASSANITO.

44 *Esso War Map* (CONVENT STATION, N.J. GENERAL DRAFTING FOR STANDARD OIL OF NEW JERSEY, [1942]). [GENERAL DRAFTING COLLECTION]

See color plate on page 92.

When the United States government rationed domestic use of gasoline and paper during the Second World War, automobile travel dwindled and most oil companies curtailed the distribution of free road maps that they had provided their customers since the 1920s. Esso (now Exxon) and its map publisher, the General Drafting Company, filled the void with a series of patriotic maps depicting the geographic background of the war. This first map in the series places the Americas at the center of the global conflict and specifies sea and air distances between strategic points. The illustrated back of the map shows the many ways in which the petroleum industry had been "drafted for the duration."

GETTYSBURG BATTLE-FIELD.

43

CONQUERING DISTANCES

45   PETRUS ROSELLI, [PORTOLAN CHART OF THE MEDITERRANEAN SEA] (MAJORCA, 1456). [AYER MS MAP 3]

The oldest-known example of a portolan chart, the remarkably accurate late-medieval charts of the Mediterranean and adjacent seas, dates from the late thirteenth century. The grid of intersecting rhumb lines represents compass bearings and may have been intended to help sailors navigate across the Mediterranean Sea beyond the sight of land. Nineteenth-century scholars applied the name "portolan" to these charts in the mistaken belief that they accompanied books of sailing directions called *portolani*. Most of the several hundred surviving examples from the late thirteenth through the early seventeenth century are, like Roselli's, handsomely colored and decorated. They were probably intended to be reference maps in the libraries of the powerful and privileged rather than practical shipboard guides. The flags that appear sporadically around the map indicate who controlled the adjacent stretch of coastline.

References: CAMPBELL 1987; SMITH, NO. 3.

46   CONTE DI OTTOMANO FREDUCCI, [MAP OF THE CENTRAL MEDITERRANEAN SEA WITH THE ADRIATIC AND AEGEAN SEAS], IN HIS [PORTOLAN ATLAS OF FIVE CHARTS] (ANCONA, 1533). [AYER MS MAP 8]

See color plate on page 92.

The nine atlases produced between 1497 and 1539 by Conte di Ottomano Freducci in Ancona—a major center of portolan chartmaking during the fifteenth and sixteenth centuries—closely resemble one another, suggesting that they were part of a substantial production from his atelier. Most portolan atlases duplicate the contents of a single-sheet portolan chart (see item 45), but in a convenient and durable format suitable both for private libraries and shipboard use. This example divides the Mediterranean and adjacent waters into five sections. The third chart in the atlas, shown here, depicts the European coast from Italy to western Turkey as well as the coasts of modern Tunisia, Libya, and western Egypt. The brilliant solid color of the many islands in the Aegean Sea and elsewhere helped to make these important navigational features more visible. Likewise, the pronounced scalloping of the coastlines typical of portolan charts may have helped mariners identify major capes and headlands.

References: AKERMAN 1995, CAMPBELL 1987.

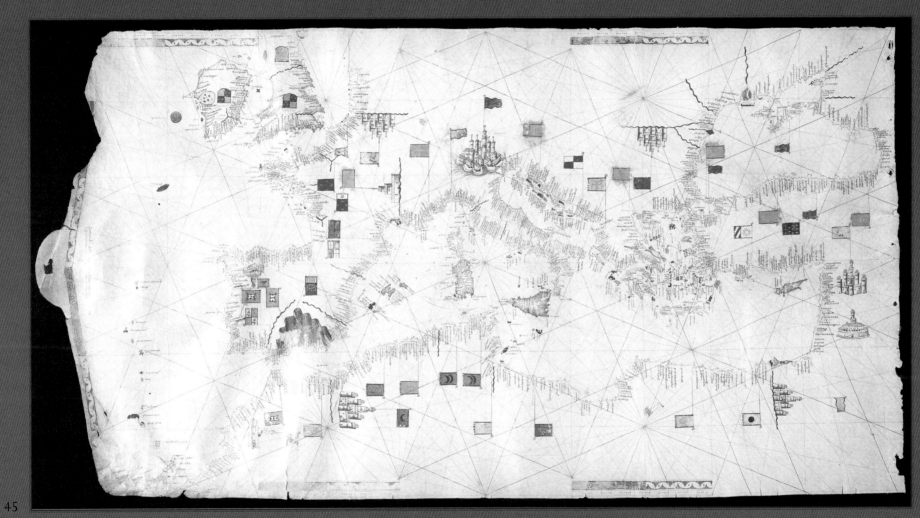

45

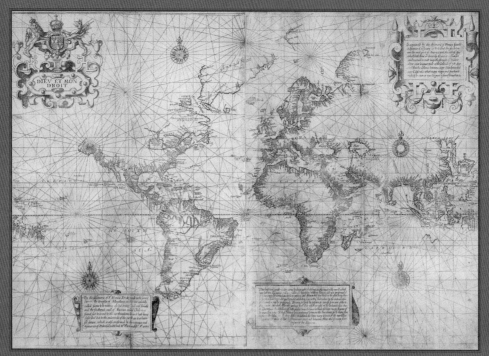

47

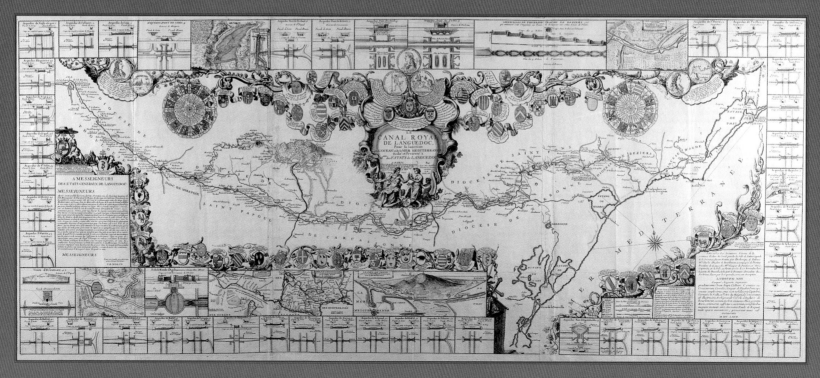

48

47 EDWARD WRIGHT, "THOU HAST HERE (GENTLE READER) A TRUE HYDROGRAPHICAL DESCRIPTION OF SO MUCH OF THE WORLD AS HATH BEEN HETHERTO DIS-COUERED AND IS COMNE TO OUR KNOWLEDGE . . . ," FROM RICHARD HAKLUYT, *The Principal Navigations, Voyages, Traffiques and Discoveries of the English Nation* (LONDON: GEORGE BISHOP, RALPH NEWBERIE, AND ROBERT BARKER, 1599). [AYER *MAP9C 0G3200 1599 .W7]

The idea of a British Empire might plausibly be dated from Hakluyt, and maps like this one, drawn on Mercator's projection, documented that tide of conquering pink. The fine detail and radiating rhumb lines of this engraving—published in the second edition of Hakluyt's famous narratives of travel and discovery—were probably the inspiration for a line in Shakespeare's *Twelfth Night:* "He does smile his face into more lines than is in the new map with the augmentation of the Indies." The only map projection to preserve true compass courses all over its surface, the Mercator remains the navigator's favorite.

References: FITE AND FREEMAN, PP. 100–102; KARROW 1993, NO. 56/17.8.

48 JEAN BAPTISTE NOLIN, *Le Canal Royal de Languedoc* (PARIS, 1697). [MAP8F 0G5832.M56 1697 .N6]

This famous inland water route, known alternately as the *canal royal,* the *canal des deux mers* (canal of the two seas), or the *canal du Midi* (canal of the south of France), connected the Bay of Biscay with the Mediterranean. One of the many public works projects sponsored by Louis XIV, it came at the beginning of the great era of canal building, which persisted well into the nineteenth century. The insets around the central map show details of the various aqueducts and locks, as well as coats of arms belonging to members of the Estates of Languedoc.

49 EDMOND HALLEY, *The Description and Uses of a New and Correct Sea-Chart of the Western and Southern Ocean Shewing the Variation of the Compass* (LONDON: MOUNT AND PAGE, 1745). [MAP6F oG9101.C93 1745 .H3]

Ancient mariners knew how to fix their position north or south of the equator by measuring the angle between the sun or stars and the horizon. Determining ships' east-west position, or longitude, however, vexed navigators for centuries, until John Harrison constructed his marvelously accurate clock in the late eighteenth century. Before Harrison, travelers held out hope for a magnetic solution based on the observation that compass needles usually do not point to true north. Edmond Halley, the astronomer royal, hoped that if he mapped the variation of the needle, navigators would be able to compare their observed variations with those shown on the chart, and thus fix their longitude. Alas, the variations Halley so painfully mapped were themselves moving targets—constantly, if slowly, changing.

References: GARDINER; ROBINSON, P. 49; THROWER.

50 GEORGE TAYLOR AND ANDREW SKINNER, "ROAD FROM EDINBURGH TO FORT STIRLING, FORT TYNEDRUM AND FORT WILLIAM," IN THEIR *Survey and Maps of the Roads of North Britain or Scotland* (LONDON: TAYLOR & SKINNER, 1776). [SC 1002]

Britons are said to have pioneered modern tourism, and they may claim the invention of the road atlas as well. Those published in Britain during the eighteenth century followed the model of a seventeenth-century prototype, John Ogilby's *Britannia* (1675). They delineated highways between major cities in a series of connecting strip maps that followed the contours of the route, in much the same manner as a modern AAA *Triptik*. This copy of Taylor and Skinner's atlas of "North Britain" retains its original soft leather binding that could fold to fit into a large pocket or saddlebag. Sheet 14, seen here, shows the first stretch of road leading northwest from Edinburgh through Stirling to Fort William, on the west coast of Scotland. The maps depict those features most essential to travelers, such as crossroads, bridges, and tollhouses, as well as major estates.

References: ADAMS; ANDREWS 1969.

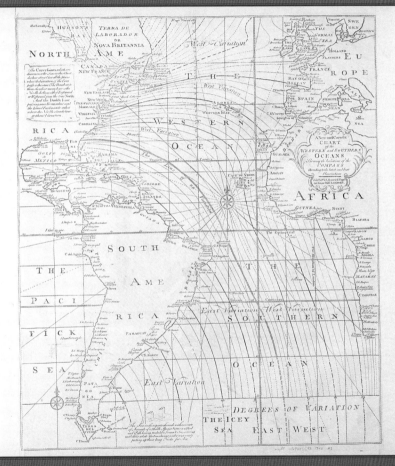

49

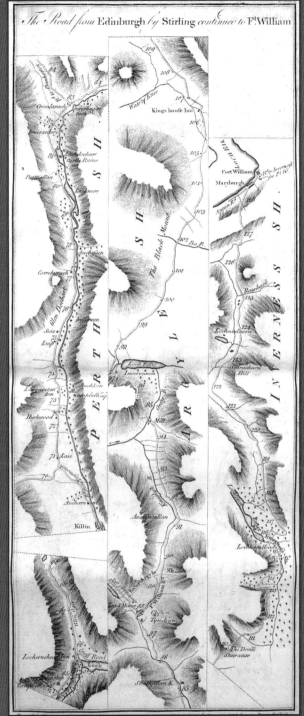

The Road from Edinburgh by Stirling continued to F.ᵗ William

50

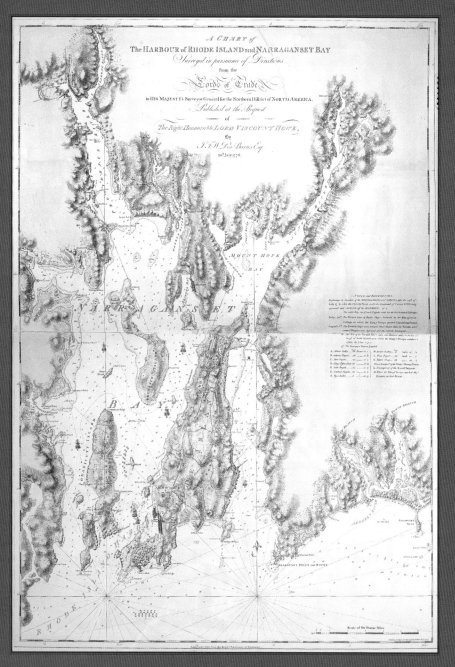

51

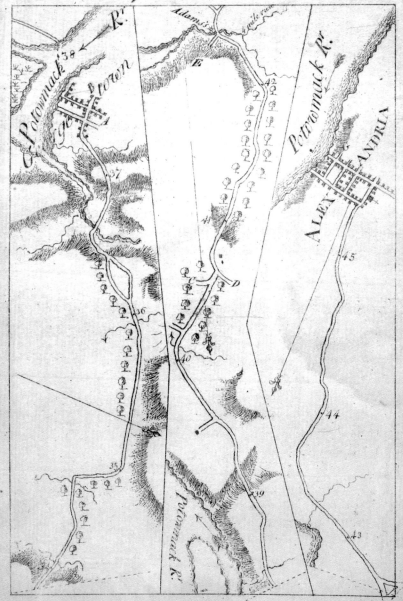

52

51 JOSEPH FREDERICK WALLET DESBARRES, "A CHART OF THE HARBOUR OF RHODE ISLAND AND NARRAGANSET BAY," FROM HIS *Atlantic Neptune* (LONDON: DESBARRES, 1777–81). [MAP4F 3320, SHEET 86F]

One of the most ambitious publishing projects of the eighteenth century, the *Atlantic Neptune* was the first systematic effort to chart American waters. It was also a curious blend of public and private enterprise: the Admiralty supplied men, ships, and a chunk of funding; DesBarres published the charts from his own home and sold them on his own account. The hydrographic surveying was finished just prior to the start of the American Revolution, and publication of the charts continued throughout the conflict. In its fullest form the Neptune contained 146 views and 115 charts and maps and was printed on almost 300 large sheets of paper. This chart was accompanied by a city plan of Newport (item 64). The cantankerous and indomitable DesBarres ended his career, at the age of 92, as lieutenant-governor of Prince Edward Island. He died in 1824, aged 103 years.

References: EVANS 1969; HARLEY *et al.* 1978, PP. 87–91; MORRISON; STEVENS.

52 CHRISTOPHER COLLES, "FROM ANNAPOLIS TO BLADENSBURG TO ALEXANDRIA," FROM HIS *A Survey of the Roads of the United States of America* ([NEW YORK: COLLES, 1789]). [*CASE RUGGLES 68]

It was no mere coincidence that Americans first published a road atlas in the same year that they ratified a constitution strengthening the political ties between states. Versatile scholar, inventor, and businessman Christopher Colles hoped that his atlas—made up of strip maps showing the main routes of the young United States—would serve as a stimulus for road improvements sorely needed to facilitate interstate communication and trade. Sheet 65, seen here, is part of a series charting the road from the Maryland state capital to Alexandria, Virginia. The section of the highway traversing the area between the Potomac River crossing at Georgetown and Alexandria passes through current-day Arlington, Virginia, and the sites of Arlington National Cemetery, the Pentagon, and Ronald Reagan Airport.

References: JOHNSON AND PETERS, PP. 26–27; RISTOW 1961.

53 ZADOK CRAMER, "MAP XI," IN HIS *The Navigator, or the Traders' Useful Guide in Navigating the Monongahela, Allegheny, Ohio, and Mississippi Rivers* (PITTSBURGH: ZADOK CRAMER, 1806). [GRAFF *5010]

Despite early efforts at federal road building, rivers and canals remained the main interior avenues of commerce in the United States during the early nineteenth century. Zadok Cramer's pocket-sized guide helped countless early travelers, traders, and boatmen find navigable channels and safe landings down the main rivers of the trans-Appalachian West. The eleventh map in the guide, shown here, illustrates the point at which the Mississippi crossed 31°north latitude and passed from the original American territory into the Louisiana Purchase lands acquired in 1803. Crudely printed from engraved wood, the map offers little detail beyond a few place names and the suggested line of navigation. The accompanying text describes landmarks and warns of dangerous obstacles, shallows, and currents.

Reference: RISTOW 1985, PP. 235–37.

54 CHARLES PREUSS, "SECTION IV," FROM HIS *Topographical Map of the Road from Missouri to Oregon . . . in VII Sections* (BALTIMORE: E. WEBER & CO., 1846). [GRAFF 3360].

Preuss based his map on notes and sketches compiled in 1843 and 1844 during John Fremont's second expedition to Oregon and California. Drawn at the relatively large scale of 10 miles to 1 inch, its extensive notes and detailed records of the Fremont party's campsites made it an invaluable guide to the Oregon Trail, then entering its period of heaviest use (in 1847 it carried four thousand travelers). Congress published and distributed the map in large numbers, but curiously, none of the narratives written by travelers on the trail seem to mention its on-road use. The section shown here includes the South Pass, where the trail crosses the Continental Divide.

References: PREUSS; STORM, NO. 3360; UNRUH; WHEAT, VOL. 3, PP. 25–29.

55 GRENVILLE MELLEN DODGE, *Map of the Military District, Kansas and the Territories. Major Gen. G. M. Dodge Commanding. 1866. Executed under the direction of Maj. Geo. T. Robinson Chf. Engr. Drawn by T. H. Williams.* (FT. LEAVENWORTH, KANS., 1866). [*MS MAP6F 0G4050 1866 D6]

See color plate on page 93.

Wounded in the Atlanta Campaign of the American Civil War, General Dodge was put in charge of the army on the western plains in 1865, where his implicit mission was to push Indian tribes north and south to make way for a central railroad route. He soon began compiling this manuscript map, completing it in early 1866. That same year, he resigned his army commission and was made chief engineer of the Union Pacific Railroad. Evidence of numerous erasures indicates that the map was revised to show new track alignments as construction of the road proceeded. Dodge said the map was "probably the best that has ever been gotten up of the country embraced in my command," and the army paid for full-sized, photographic reproductions to issue to western officers. The Newberry collections include two copies of these early "photocopies."

References: PARKE-BERNET; THURMAN.

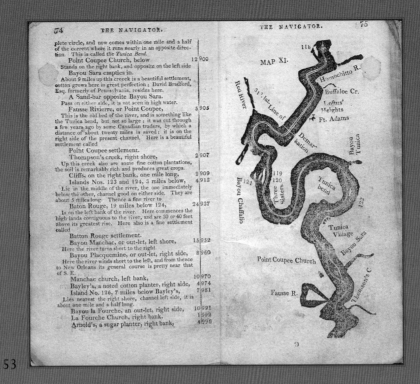

plete circle, and now comes within one mile and a half of the current where it runs nearly in an opposite direction. This is called the *Tunica Bend*.

Point Coupee Church, below — 12 900

Stands on the right bank, and opposite on the left side

Bayou Sara empties in.

About 9 miles up this creek is a beautiful settlement, cotton grows here in great perfection; David Bradford, Esq. formerly of Pennsylvania, resides here.

A Sand-bar opposite Bayou Sara.

Pass on either side, it is not seen in high water.

Fausse Rivierre, or Point Coupee, — 5 905

This is the old bed of the river, and is something like the Tunica bend, but not so large; it was cut through a few years ago by some Canadian traders, by which a distance of about twenty miles is saved; it is on the right side of the present channel. Here is a beautiful settlement called

Point Coupee settlement.

Thompson's creek, right shore, — 2 907

Up this creek also are some fine cotton plantations, the soil is remarkably rich and produces great crops.

Cliffs, on the right bank, one mile long, — 2 909

Islands Nos. 123 and 124, 3 miles below, — 4 913

Lie in the middle of the river, the one immediately below the other, channel good on either side. They are about 5 miles long. Thence a fine river to

Baton Rouge, 19 miles below 124, — 24 937

Is on the left bank of the river. Here commences the high lands contiguous to the river, and are 30 or 40 feet above its greatest rise. Here also is a fine settlement called

Batton Rouge settlement.

Bayou Manchac, or out-let, left shore, — 15 952

Here the river turns short to the right.

Bayou Placquemine, or out-let, right side, — 8 960

Here the river winds short to the left, and from thence to New Orleans its general course is pretty near that of S. E.

Manchac church, left bank, — 10 970

Bayley's, a noted cotton planter, right side, — 4 974

Island No. 126, 7 miles below Bayley's, — 7 981

Lies nearest the right shore, channel left side, it is about one mile and a half long.

Bayou la Fourche, an out-let, right side, — 10 991

La Fourche Church, right bank, — 1 992

Arnold's, a sugar planter, right bank, — 4 996

MAP XI.

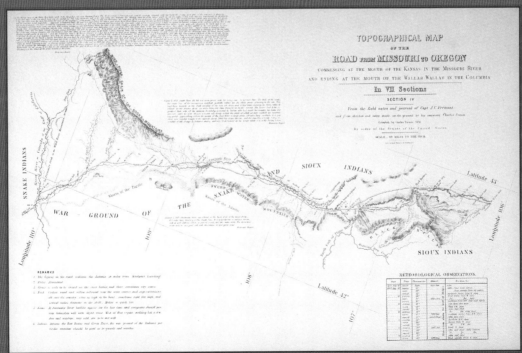

TOPOGRAPHICAL MAP
OF THE
ROAD FROM MISSOURI TO OREGON
COMMENCING AT THE MOUTH OF THE KANSAS IN THE MISSOURI RIVER
AND ENDING AT THE MOUTH OF THE WALLAH WALLAH IN THE COLUMBIA

In VII Sections

SECTION IV

From the field notes and journal of Capt. J. C. Fremont
and from sketches and notes made on the ground by his assistant Charles Preuss

Compiled by Charles Preuss, 1846

By order of the Senate of the United States

SCALE, 10 MILES TO THE INCH.

METEOROLOGICAL OBSERVATIONS.

CHICAGO TO MILWAUKEE
No. 18

61.6 Miles
Here right-hand road leads to Kenosha. Left-hand road to Lake Geneva. Continue straight ahead for Milwaukee for 6.4 miles.

No. 19

68.0 Miles
Here keep straight ahead for Milwaukee to end of road five-tenths of a mile. Right-hand road leads to Racine.

CHICAGO TO MILWAUKEE
No. 20

68.5 Miles
Turn left (west) and keep straight ahead to Kilbourn Road, 3.8 miles.

No. 21

72.3 Miles
Turn right (north) on Kilbourn Road and continue straight ahead paying no attention to cross roads for 22.4 miles to Forest Home Avenue, Milwaukee.

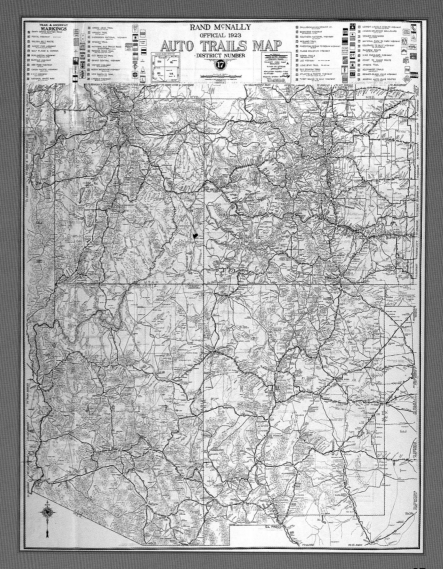

RAND McNALLY
OFFICIAL 1923
AUTO TRAILS MAP
DISTRICT NUMBER 17

56 *The Rand McNally Photo-Auto Guide, Chicago to Milwaukee / Milwaukee to Chicago* (CHICAGO: RAND MCNALLY, 1909). [RAND MCNALLY COLLECTION]

In the days before American automobile highways were systematically marked and mapped, ambitious and cartographically inclined motoring enthusiasts found that their knowledge of specific routes could be a marketable commodity. Andrew McNally II, son of the cofounder of Rand McNally, was so immersed in his family's business that, legend has it, he paused periodically during his honeymoon trip to take these photographs. They later became the basis for the Chicago-to-Milwaukee volume in the company's series of "photo-auto guides." A photograph of each major crossroads or fork in the road would ensure that motorists made the correct turn. The photographs were keyed to a series of maps of the entire route. H. Sargent Michaels developed the concept for these route guides in 1905. Rand McNally took over publication of the series in 1909.

References: AKERMAN 1993A; RISTOW 1946.

57 *Rand McNally Official 1923 Auto Trails Map, District No. 17* (CHICAGO: RAND MCNALLY, 1923). [RAND MCNALLY COLLECTION]

Rand McNally's reputation as a publisher of road maps was established by its popular "auto trails" map series, published from 1917 to the mid-1920s. Throughout this period, most American trunk routes were identified by names invented by motor clubs and highway boosters, rather than by numbers. This chaotic and informal system was a nightmare for cartographers, who struggled to find a convenient way to identify the routes without crowding out other essential information. Rand McNally cartographer John Garrett Brink's ingenious solution was to assign a number to each route name. The advertisements on the Auto Trails maps promoted local roadside businesses that paid for advance subscriptions. This marketing strategy laid the groundwork for widespread distribution of cheap road maps by oil companies and related businesses that would become the norm from the late 1920s through the 1970s.

References: AKERMAN 1993B, RISTOW 1946.

58 ERHARD REEUWICH, "CIVITAS VENETIARU[M]," FROM BERNHARD VAN BREYDENBACH, *Sanctarum peregrinationum in montem Syon ad venerandum Christi sepulchrum in Jerusalem* (MAINZ: PETER SCHÖFFER, 1486). [NOVACCO 9F 3]
See color plate on pages 94–95.

This first printed collection of city views was based on records of a pilgrimage to the Holy Land. The pilgrim, a wealthy nobleman named Bernhard van Breydenbach, brought the artist Erhard Reeuwich on his journey to document the places he visited. The published account of his travels includes several woodcuts made from Reeuwich's drawings. An important staging point on the way to the Holy Land, Venice inspired this splendid view from the perspective of an approaching ship. This type of view, drawn from a point at or close to ground level, is called a profile. The illustration would have to have been folded many times to fit into the folio volume.

References: SCHULZ 1970, PP. 30–31, 93; TALBOT.

59 JACOPO DE BARBARI, *Venetie MD* (VENICE: [BARBARI], 1500). [NOVACCO *6F 7]

Few city portraits can equal the grandeur of Barbari's Venice, the largest and most magnificent view of its day, created to glorify the city and advertise its primacy as a center of trade. The figure of Mercury, symbol of commerce, looms at top center with the inscription "I Mercury shine favorably on this above all other emporia." Below the city, the sea-god sits astride a great fish, holding a banner that reads "I Neptune reside here, smoothing the waters at this port." Observed from an imaginary viewpoint high above the ground, Barbari's vision of Venice can aptly be called a bird's-eye view.

Reference: SCHULZ 1978.

60 HERNANDO CORTES, [MAP OF TENOCHTITLAN AND THE GULF OF MEXICO], IN HIS *Praeclara Ferdinãdi Cortesii de Nova maris Oceani Hyspania Narratio . . .* (NUREMBERG: F. PEYPUS, 1524). [AYER *655.51 C8 1524D]
See color plate on page 94.

To the *conquistadores* sweating their way upcountry from Veracruz, the first sight of the Aztec capital Tenochtitlan must have been a shock. Bernal Diaz said it looked "like the enchantments they tell of in the legend of Amadis [in the best-selling romance of the era], on account of the great towers and temples and buildings rising from the water." In May of 1522, Cortes sent the drawing on which this woodcut was based to Charles V. Recent scholarship has argued convincingly that the drawing was copied, at least in part, from an indigenous Aztec map. The map shown here provided Europeans with their first glimpse of a major non-European city, complete with temples, residences, gardens, causeways, and a zoo. As an "island" city, Tenochtitlan (centered on the Plaza Mejor in present-day Mexico City) was soon reproduced in the form of *isolari,* as well as in the Braun and Hogenberg book of town views (item 61).

References: HARLEY 1990, PP. 79–81; MUNDY 1998.

59

61 "Plona," in Georg Braun and Frans Hogenberg, *Civitates orbis terrarum,* 6 vols. (Cologne: G. Kempensem, 1572–1617). [Ayer *+135 B8 1573]
See color plate on page 95.

Braun and Hogenberg envisioned their massive collection of urban plans and views as a companion to Abraham Ortelius's world atlas, *Theatrum orbis terrarum.* Many of the maps in the *Civitates* had been published previously, but never before had such a complete view of urban life around the world been offered between the covers of a book. This view shows the relatively small town of Plön, a provincial capital in the Duchy of Holstein, then a fief of the Danish crown, now part of the German state of Schleswig-Holstein. It derives from a manuscript drawing sent to Braun by the Danish statesman and scholar Heinrich Rantzau. The idyllic mood set by images of fishermen on the chain of lakes surrounding the city is broken somewhat by the lakeside gallows at left and the two fortresses (each labeled "arx") that dominate the town and country.

References: Koeman 1967–85, vol. 2, pp. 9–25; Skelton 1965–66.

62 Jan Jansson, "Lugdunum vulgo Lyon," in *Illustriorum regni Galliæ civitatum tabulæ . . . ,* vol. 4 of his [Theatrum Urbium] (Amsterdam: Jansson, 1657). [Case +G 117.452, vol. 4]
See color plate on page 96.

Lyon, at the confluence of the Saône and Rhône Rivers, has long been called France's "second city." The capital of Roman France, it grew from an important regional hub into a major center of commerce in the Middle Ages, when it began hosting important semi-annual, and later, quadrennial fairs. Jansson's view represents a reduced version of a large three-sheet view by the Lyon surveyor Simon Maupin that was originally published in 1625. The city, enclosed in its medieval walls, boasted a great cathedral and a number of smaller churches, all identified by number in the legend. The legend also identifies a small cluster of pesthouses, or shelters for plague victims, outside the walls at the left side of the view.

References: Koeman 1967–85, vol. 2, pp. 189–204; Lyon.

63 Louis Bretez, *Plan de Paris* ([Paris: 1739]). [Wing *+ZP 739.B844]

See overleaf.

Often called the Turgot plan after its commissioner Michel-Etienne Turgot, the map shown here reflects the same brand of civic pride and self-promotion seen in the Barbari plan of Venice (item 59). Turgot himself held the high municipal office of Prévôt des Marchands. Like the Barbari, Bretez's 20-sheet plan was a monumental production. Sheet 11, shown here, includes a large portion of the ancient city center, the Isle de la Cité, dominated at top left by Notre Dame and at bottom left by the Pont Neuf. To the right, at some distance from the river, are the grounds of Luxembourg Palace, built in 1615 for Marie de Médicis. In the era of the Bretez plan, many of the city's leading map publishers and printers could be found on the Quai de l'Horloge and other streets on the Isle de la Cité.

Reference: Miller; Pedley; Rouleau.

64 Charles Blaskowitz, "A Plan of the Town of Newport in the Province of Rhode Island" (1776), in J. F. W. DesBarres, *Atlantic Neptune* (London: DesBarres, 1774–81). [Map4F 3320, sht. 87c]

See overleaf.

Although most of the maps included in DesBarres's massive collection are navigational charts of American waters, a few cities warranted separate street plans, among them Halifax, Boston, Charleston, and Newport. As one of the richest and most flourishing cities on the continent, Newport was a logical choice. Home to more than nine thousand people in 1775, it was the fifth largest American city, trailing only Philadelphia, New York, Boston, and Charleston. DesBarres's engraving, based on a survey by Charles Blaskowitz, caught Newport at the high tide of its eighteenth-century prosperity. Soon after, the British army occupied the city, raiding its wharves for firewood and prompting many of its merchants and citizens to flee.

References: Cappon, pp. 81–82, 97.

63 (detail)

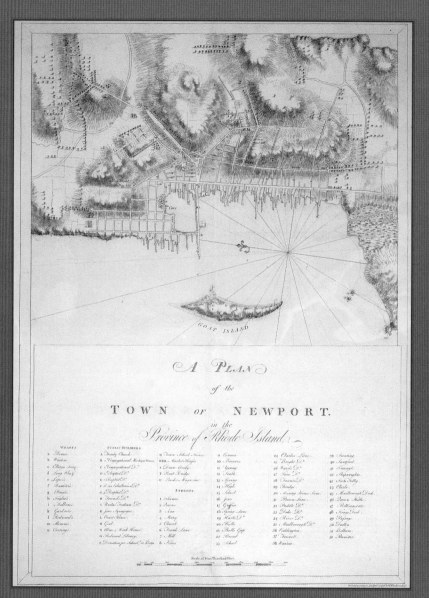

64

65   Luigi Rossini, *Panorama di Roma antica e moderna* (Rome, 1827). [Novacco 8F 5]

See overleaf.

An architect who specialized in engraving Roman scenes, Luigi Rossini here portrays his city from the vantage point of the campanile of San Francesca Romana. He endeavored to show not only the ancient city, but also those modern sites "likely to be of interest to lovers of antiquities and tourists." This stunning view was printed from four large copperplates; the left and right edges could be joined to form a complete 360-degree panorama with a diameter of about 3½ feet. A good deal of archaeological work was done in Rome in the early nineteenth century, and Rossini's rendition of the Colosseum, at the far left, shows some of the modern restoration.

References: Cavazzi, pp. 106–109; Hyde.

66   *A Map of Chicago's Gangland from Authentic Sources* (Chicago: Bruce-Roberts, 1931). [Map G 10896.548]

See color plate on page 97.

Chicago's reputation as a Prohibition-era haven for gangsters was promoted by popular literature, film, and (apparently) maps. This grim but tongue-in-cheek map for tourists, "designed to inculcate the most important principles of piety and virtue in young persons and graphically portray the evil and sin of large cities," gives new meaning to our theme "celebrating the city." Highlights include the site of the St. Valentine's Day massacre, Al Capone's South Side headquarters, the county morgue, and, somewhat incongruously, the future site of the 1933 World's Fair. Skulls and crossbones mark the location of reputed gangland killings. Promotional text on the wrapper accompanying the map alleges, we hope in jest, that the authors were run out of town for their efforts.

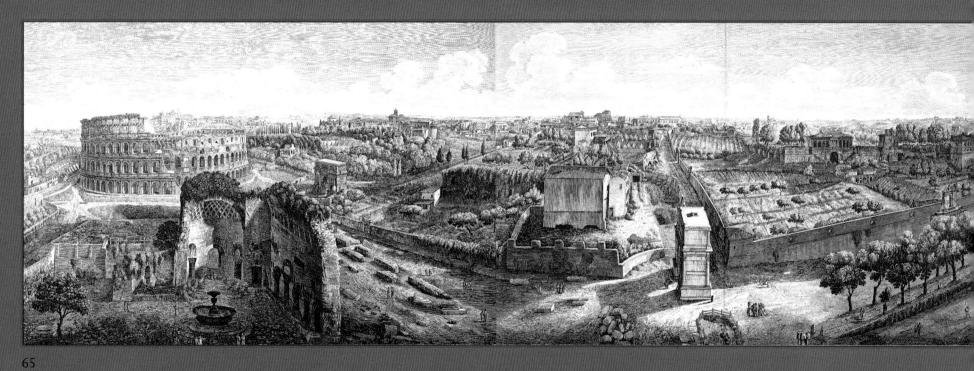

65

PLOTTING THE COUNTRYSIDE

67 [MAP OF THE TEMASCALTEPEC REGION, MEXICO] (1560s?). [AYER MS 1906]
See color plate on page 96.

One of the first regions in Mexico exploited for its mineral resources, Temascaltepec, in the mountains southwest of Mexico City, remained an important mining center throughout the colonial period. This plan, drawn on traditional Mexican "paper" made from the maguey plant, accompanied the records of a land dispute. Probably dating from the 1560s, it shows the locations of several haciendas, mines, and a smelter. Like most early Mexican maps, it is an amalgam of Spanish and native Aztec pictographic conventions. Horseshoes are an Aztec symbol for roads, the temple glyph identifies villages, and rivers are denoted with the typical Aztec spiral indicating flowing water.

References: COBARRUBIAS; GERHARD; MUNDY 1996.

68 JAN SPRUYTENBURGH, "NIEUWE KAART VAN MYNDEN EN DE LOOSDRECHT" (1734) IN REINER AND JOSHUA OTTENS, *Atlas* (AMSTERDAM: OTTENS, 1665–1750). [CASE *+0G1860.O88.A84 1750, V. 3]
See color plate on page 98.

In much of the Netherlands, and often with human help, the balance between land and water has shifted frequently. In the area north of Utrecht, shown here, small ponds formed when areas from which peat was excavated in the fourteenth century filled with water. Such pools, marshlands, and other regularly flooded areas were often pumped dry by windmills to create "polders," or land for cultivation. To track this ebb and flow, Dutch polder authorities produced dozens of maps of reclamation projects. Many are large and, like the one seen here, splendidly colored. The exquisitely engraved and colored scene at the bottom of the map adds an impressionistic view of this watery agricultural landscape.

References: KAIN AND BAIGENT; LAMBERT; RISTOW 1974.

69 "Partie de la Novvelle Angleterre" (ca. 1720), from the [Cartes Marines] (Paris?, ca. 1726). [Ayer MS Map 30, no. 88]

See color plate on page 99.

Seemingly the depiction of a peaceful New England landscape, this map actually records dark and bloody ground. In a series of raids in the early eighteenth century, the French and their Indian allies attacked the English settlements on the Massachusetts and New Hampshire frontier. The information represented here, from a manuscript atlas documenting areas of interest to the French around the world, betrays its military intent by listing the number of houses belonging to each settlement and detailing settlement defenses and numbers of armed men. There is a direct allusion to the raid on Haverhill, Massachusetts, on 29 August 1708, and a note about a barricade being constructed in Boston in 1718. It may well have been the work of a French spy; note 29 refers, with French spellings, to the "route from Vesfield [Westfield, Mass.] to Albanie [Albany, N.Y.] which the author followed on foot."

References: Buisseret 1985; Chase; Douville; Russ.

70 "Carte du Forte Rozalie de Nachez Francois avec ses dependances et village de savages" (ca. 1727), from Jean-Benjamin Francois Dumont de Montigny, *Memoire de Lxx Dxx Officiere Ingenieur, contenant les evenemens qui se sont passés à la Louisiane depuis 1715 jusqu'a present* (1747). [Ayer MS 257, map 9]
See overleaf.

In 1716, French settlers built Fort Rosalie at the site of present-day Natchez, Mississippi. Ten years later, when settlement around the fort had grown to include about a thousand farmers raising tobacco, wheat, and other crops, Sublieutenant Dumont de Montigny made this bucolic sketch of the area. The peaceful scene was short-lived. On 28 November 1729, the Natchez Indians, whose village is shown at the top center of the map, took up arms against the French. Fort Rosalie and its small garrison proved no match for them—though it may look substantial, a contemporary account called the fort "an enclosure of poor piles, half rotten, that permit free entrance almost everywhere." The Natchez destroyed the fort and nearby farms and killed or captured most of the residents. A rich source for this early period in Louisiana, Dumont de Montigny's account includes plans of New Orleans, Biloxi, and Pensacola among a total of 14 maps.

References: Buisseret 1991, pp. 94–95; Wilson.

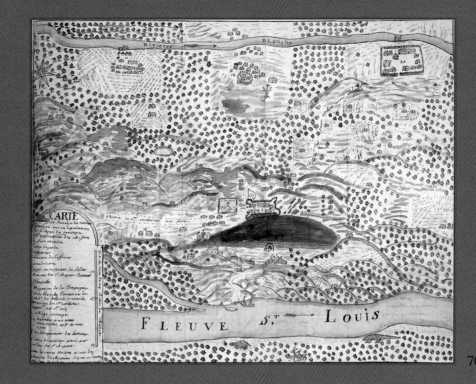

70

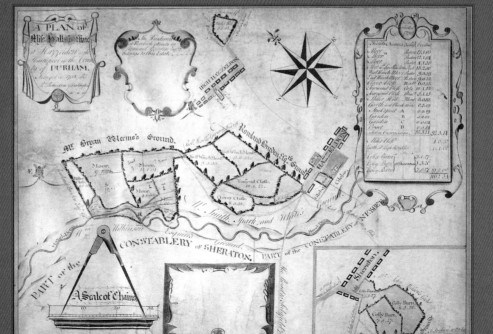

71

71 RICHARD RICHARDSON, *A Plan of Miss Hall's Estate of Hazzledon nigh Hartlepool in the County of Durham* (DURHAM?, 1758). [MS MAP8C OG5754.H47 1758]

Popular among the English aristocracy in the sixteenth century, estate plans prepared and drawn by private surveyors helped landlords manage their estates and, judging by their beautiful execution, displayed the landlord's wealth and status as well. Such maps provided an accurate account of the disposition, character, and size of lands dedicated to particular uses. Miss Hall's estate, for instance, had six fields devoted to pasture, three meadowlands, and one field each devoted to the cultivation of oats and wheat. The unit of measure on the plan indicated by the scale at lower left is the length of a surveyor's chain, equal to 22 yards.

References: BENDALL; BUISSERET 1988 AND 1996.

72 HUMPHRY REPTON, *Sketches and Hints on Landscape Gardening* (LONDON: W. BULMER, 1794). [CASE fW12.737]

See overleaf.

In reaction to geometric and stylized gardens like those at Versailles and Hampton Court, many eighteenth-century Britons discovered a taste for free-flowing "natural" landscapes. Pioneer landscape gardener Humphry Repton, who transformed parklands by relocating ponds, trees, plantings, and meadows to create picturesque rural scenes, played a central role in shaping this aesthetic. Repton's romantic style influenced such American landscape designers as Frederick Law Olmsted, who created New York's Central Park, and Jens Jensen, whose "natural prairie" design graces Chicago's Columbus Park. Repton presented his clients with plans in the form of manuscript books; paper flaps could be flipped to show watercolor renderings of the landscape before and after its transformation. *Sketches and Hints,* Repton's first published work, replicates one of those manuscripts.

References: DANIELS; LAURIE.

73 T. WHITTEN, "SPARTANBURG DISTRICT" (1820), IN ROBERT MILLS, *Atlas of the State of South Carolina* (BALTIMORE: F. LUCAS JR., 1826). [AYER +135 M6 1825]

As the first atlas devoted to a single American state, Mills's atlas set a high standard for those that followed. To fund its production, South Carolina's progressive legislature hired 19 different surveyors to map the state's 28 districts at a total cost of $90,000 (about $1.3 million in today's currency). The district surveys were then combined and reduced in scale to produce a complete map in 1822. The next year, the legislature authorized architect and engineer Robert Mills to publish the original district maps in the form of an atlas, which appeared in 1826. Nineteenth-century South Carolina was a predominantly rural state, home to large cotton and indigo plantations. Landowners' names are recorded here in a style that would later be adopted by county atlases in the 1860s and beyond.

References: RISTOW 1977; WPA.

74 JOHN J. PRATT AND BELA S. BUELL, *Map of the Gold Regions in the Vicinity of Central City, Gilpin Co., Colorado Territory* (PHILADELPHIA: W. H. REASE, 1862). [GRAFF *3344]

See overleaf.

Mineral wealth, particularly gold and silver, propelled Colorado's early growth. Pratt and Buell's map captures the atmosphere of the most productive of Colorado's gold regions, the Central City district, about 30 miles west of Denver. John H. Gregory discovered gold here in 1859; it was mined by dredging the narrow gulches or by quarrying an unusually soft form of quartz rich with the precious metal. The many quartz mills shown on the map crushed the rock and separated the gold. Alongside their map, Pratt and Buell advertised the beauty of the region's views and the civility of its towns, with an eye to attracting new settlers. New arrivals found the region already organized into mining districts (marked by hand on the map) with rules for establishing claims and informal courts for settling disputes.

Reference: STORM, NO. 3344.

72

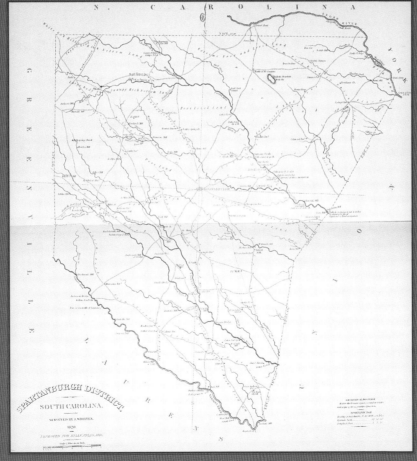

73

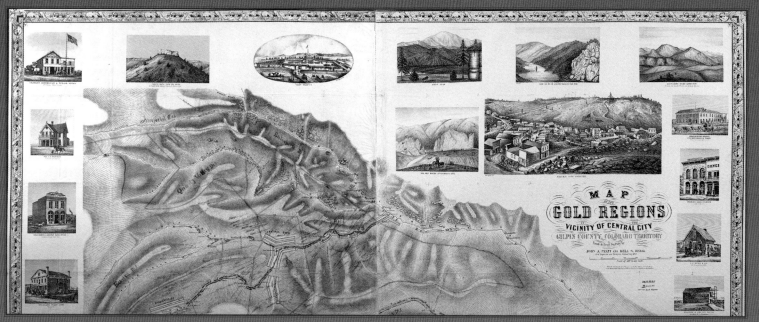

74 *(detail)*

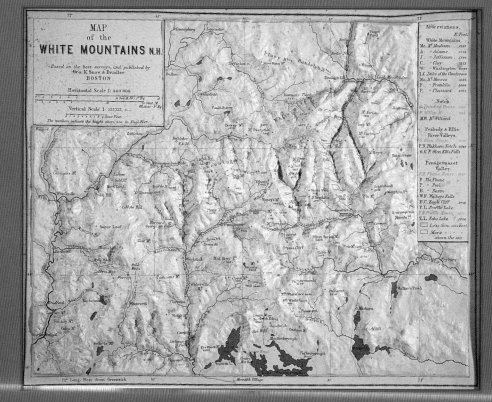

75

75 GEORGE K. SNOW & BRADLEE, *Map of the White Mountains, N.H.* (BOSTON: GEORGE K. SNOW & BRADLEE, 1872). [MAP4C oG3742.W5 1872 .G4]

An early destination for American tourists, New Hampshire's rugged White Mountains were among the best-mapped rural districts of the eastern United States in the latter half of the nineteenth century. Maps designed to attract tourists to the area or to guide them when they got there advertised the area's most spectacular assets—its steep ranges of peaks and deep, glacially carved valleys provided some of the most dramatic topography in the United States. This three-dimensional relief map offered an unusually realistic impression of the terrain just as railroads opened the region to middle-class tourists from the great urban centers of the Northeast. Mount Washington's 6,288-foot summit is visible at upper right. The highest peak in the eastern United States, it was accessible via cog railway beginning in 1869.

76 "PICTORIAL MAP OF THE 130 ACRE FARM OF JONATHAN MILLER," FROM W. R. BRINK, *Illustrated Atlas Map of Menard County, Illinois* (EDWARDSVILLE, ILL.: W. R. BRINK & CO., 1874). [+F896565.43]
See overleaf.

From the 1840s through the early twentieth century, commercial publishers compiled and published several thousand maps and atlases of American counties showing the location, size, and owners of every parcel of rural land. These landownership maps and atlases were financed in part through advance subscriptions sold to farmers and "leading citizens." For an additional fee, subscribers could have their biography, portrait, or a view of their farmstead included in the atlas. The view and map of Jonathan Miller's land shown here offer a detailed glimpse into the inner workings of a stock farm. Note Miller's convenient access to the local rail station and his proximity to the district's one-room schoolhouse.

References: CONZEN 1984, 1997.

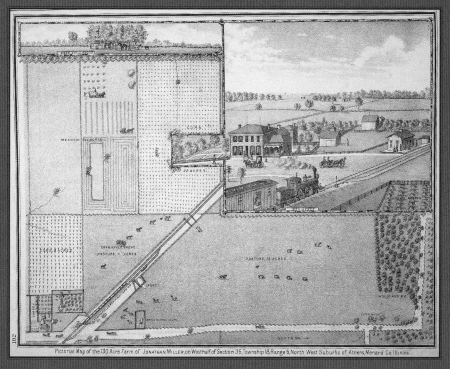

Pictorial Map of the 130 Acre Farm of Jonathan Miller, on Westhalf of Section 36, Township 18, Range 6, North West Suburbs of Athens, Menard Co. Illinois

76

77

77 TEXAS AND PACIFIC RAILWAY CO., *Map of the Great Southwest* (ST. LOUIS: WOODWARD & TIERNAN FOR THE TEXAS & PACIFIC RAILWAY CO., 1882). [CHICAGO, BURLINGTON & QUINCY ARCHIVES]

Beginning in the 1850s, federal land grants made to railroad companies helped finance the construction of tracks throughout the American West while simultaneously encouraging the settlement and development of adjacent lands. Railroad-created land companies, charged with attracting prospective settlers, produced maps that emphasized connections between existing lines, many of which were controlled by the same small group of investors. The map and booklet shown here, published by the land department of the Texas and Pacific Railway in 1882, underscored the links between its own tracks and those of the Missouri Pacific, the Union Pacific, and other cooperating lines (dubbed the Southwestern Railway System). Note the illustrations of a special sleeper car for "emigrants" and an "immigrant house" built by the railroad in Baird to house new arrivals.

Reference: MODELSKI 1984.

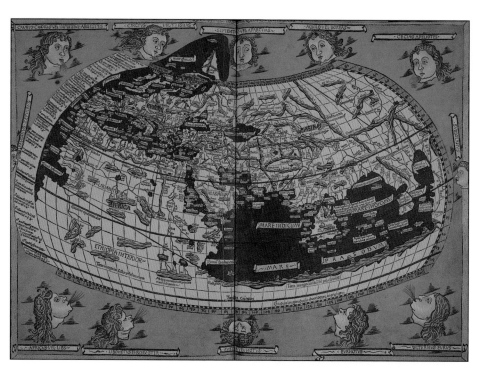

2 CLAUDIUS PTOLEMY,
[WORLD MAP] (1482).
See caption on page 14.

6 BAPTISTA AGNESE,
[PORTOLAN ATLAS]
(CA. 1550).
See caption on page 18.

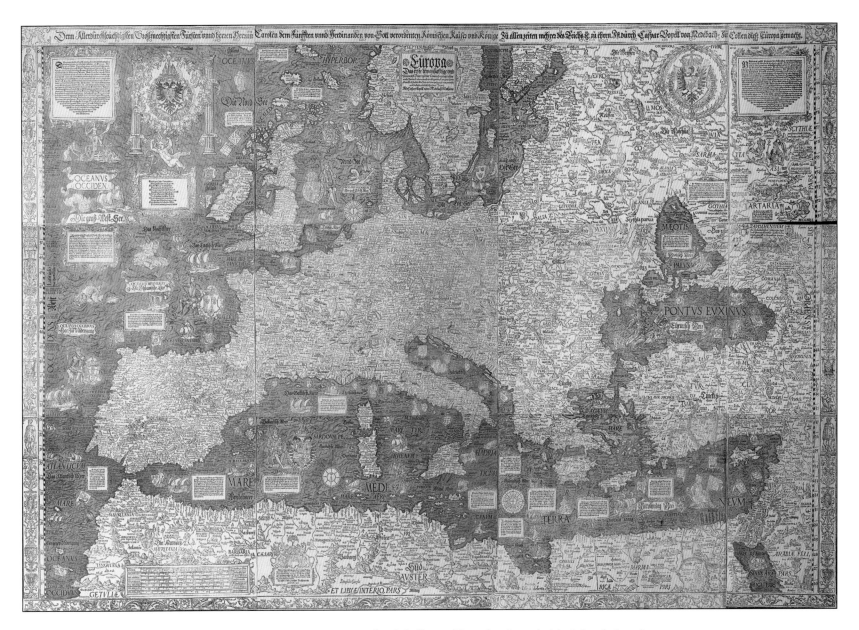

10 KASPAR VOPEL, *Europa. Das erste bewohnhafftige und fürnembste Dritte theil des Erdtreichs* (1597).
See caption on page 22.

NOVA TOTIVS TERRARVM ORBIS TABVLA

13 Frederik de Wit and Giuseppe Longhi, *Nova totius terrarum orbis tabula* (CA. 1680).
See caption on page 25.

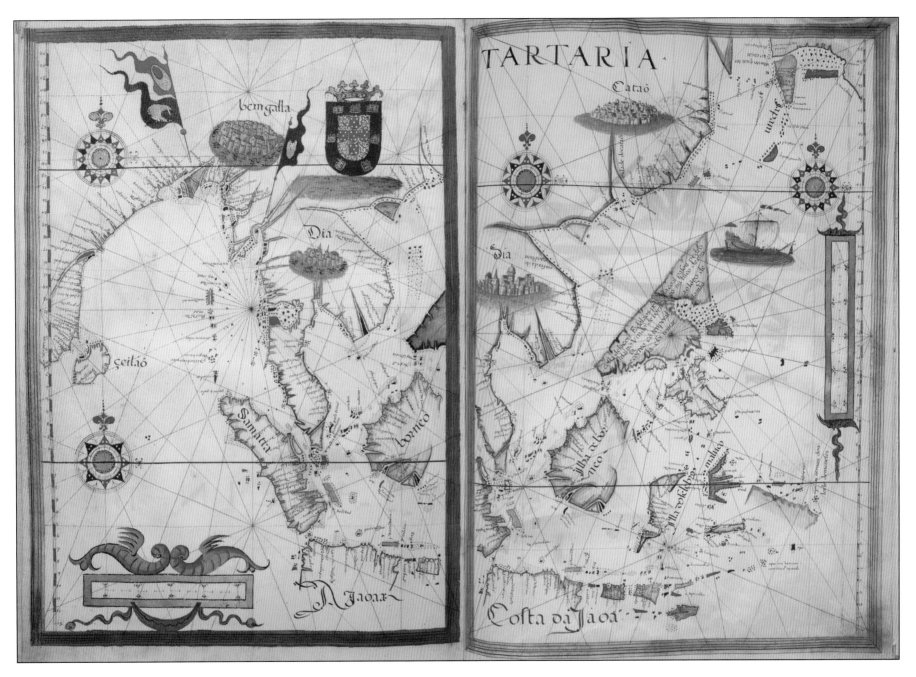

20 Sebastião Lopes, [Charts of the Bay of Bengal and Southeast Asia] (ca. 1565).
See caption on page 32.

22 CHRISTOPHER SAXTON, "NORTHAMTON, BEDFORDIAE, CANTABRIGIAE, HUNTINDONIAE, ET RUTLANDIAE COMITATUUM . . . 1576" (1579).

See caption on page 32.

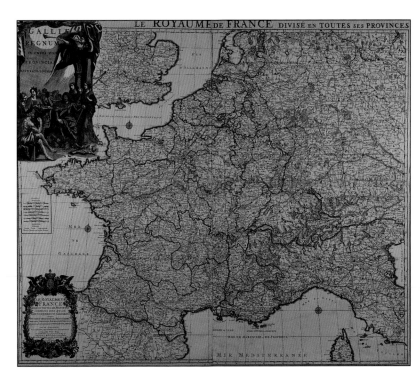

26 Alexis Hubert
Jaillot, *Galliae Regnum.*
Le Royaume de France
(ca. 1725).
See caption on page 36.

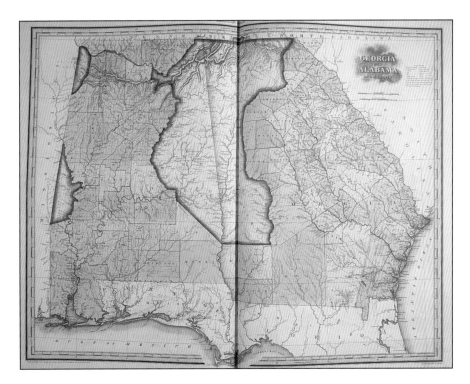

29 Henry S. Tanner, "Georgia
and Alabama" (1823).
See caption on page 39.

31 John Disturnell,
*Mapa de los Estados Unidos
de Méjico* (1847).
See caption on page 40.

33 William Henry
Holmes, "The Grand
Cañon at the Foot of
the Toroweap —
Looking East" (1882).
See caption on page 43.

THE GRAND CAÑON AT THE FOOT OF THE TOROWEAP·LOOKING EAST

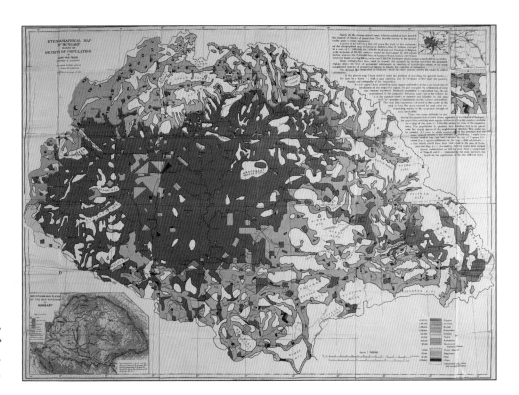

34 Count Pál Teleki,
"Ethnographical Map
of Hungary" (1923).

See caption on page 43.

39 John Mitchell,
*A Map of the British and
French Dominions in
North America* (1755).

See caption on page 48.

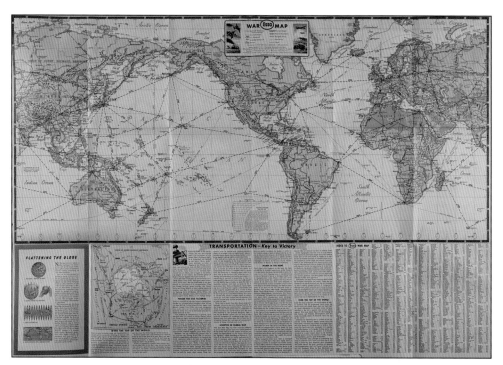

44 *Esso War Map* [1942].
See caption on page 52.

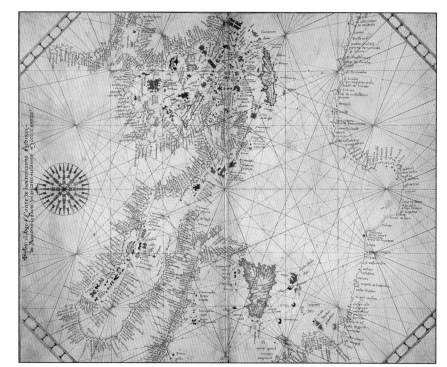

46 CONTE DI OTTOMANO
FREDUCCI, [MAP
OF THE CENTRAL
MEDITERRANEAN SEA
WITH THE ADRIATIC AND
AEGEAN SEAS] (1533).
See caption on page 54.

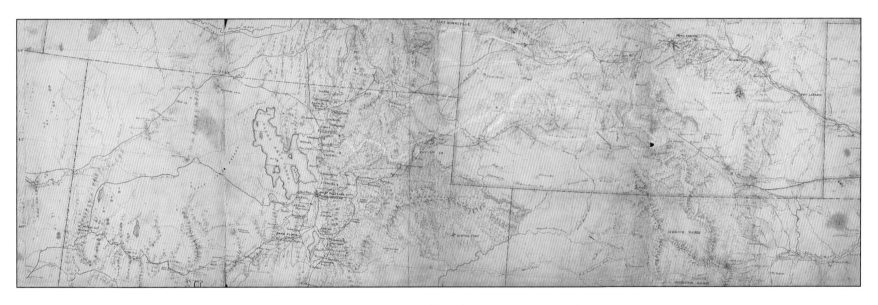

55 *(detail)* GRENVILLE MELLEN DODGE, *Map of the Military District, Kansas and the Territories* (1866).
See caption on page 62.

58 ERHARD REEUWICH,
"CIVITAS VENETIARU[M]" (1486).
See caption on page 66.

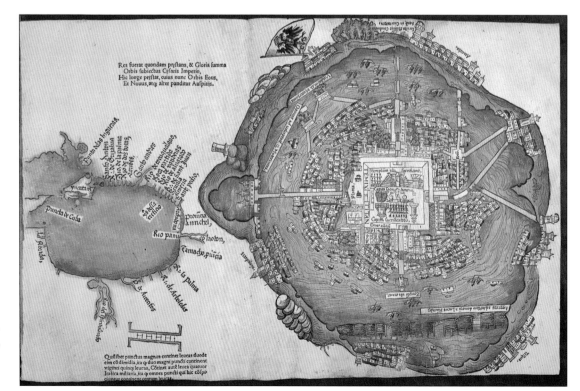

60 HERNANDO CORTES,
[MAP OF TENOCHTITLAN
AND THE GULF OF
MEXICO] (1524).
See caption on page 66.

61 "PLONA" (1572–1617).

See caption on page 68.

95

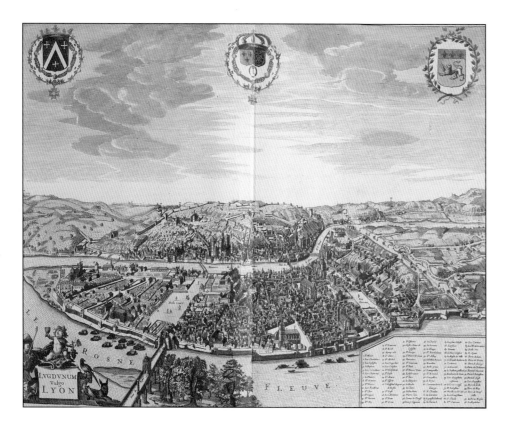

62 JAN JANSSON,
"LUGDUNUM VULGO
LYON" (1657).
See caption on page 68.

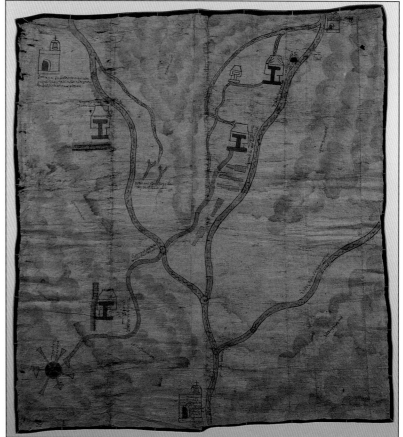

67 [MAP OF THE
TEMASCALTEPEC REGION,
MEXICO] (1560s?).
See caption on page 74.

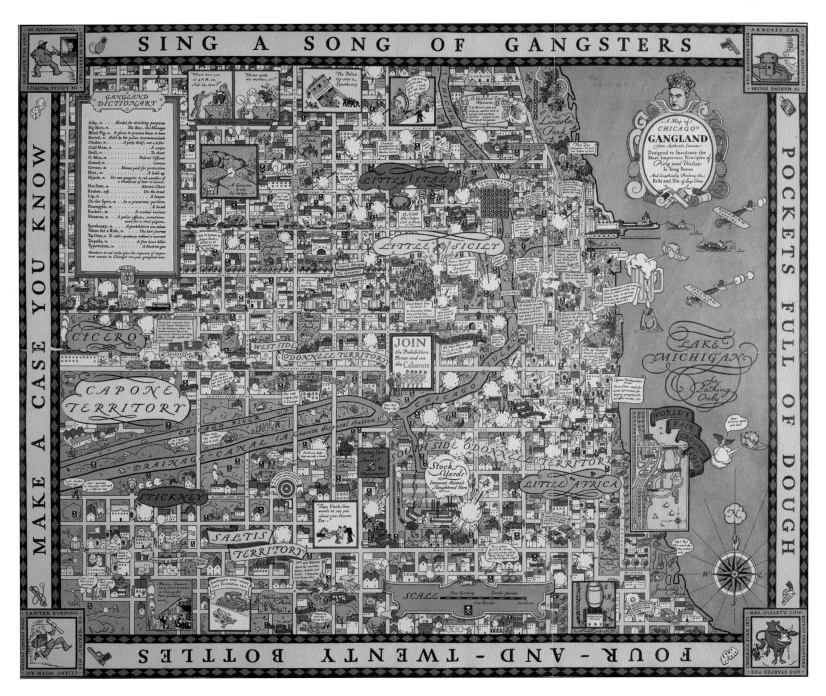

66 *A Map of Chicago's Gangland from Authentic Sources* (1931).
See caption on page 71.

68 Jan Spruytenburgh, "Nieuwe kaart van Mynden en de Loosdrecht" (1734).
See caption on page 74.

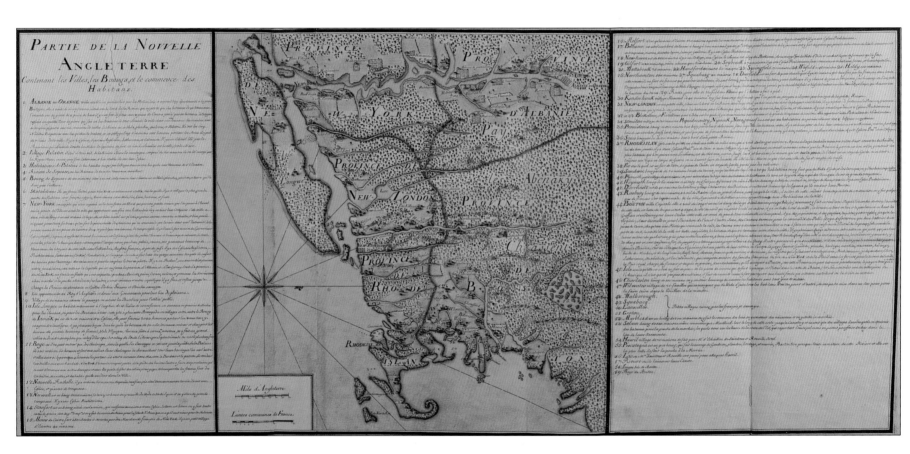

69 "Partie de la Nouvelle Angleterre" (ca. 1720).

See caption on page 75.

BIBLIOGRAPHY

Adams, I. H. 1975. "George Taylor, a Surveyor o' Parts," *Imago mundi* 27:65–72.

Akerman, James. 1991–92. "A Collection of Haas Typographic Maps," *Mapline* 64:1–5.

———. 1993a. "Blazing a Well-worn Path: Cartographic Commercialism, Highway Promotion, and Automobile Tourism in the United States, 1880–1930," *Cartographica* 30, no. 1:10–20.

———. 1993b. "Selling Maps, Selling Highways: Rand McNally's 'Blazed Trails' Program," *Imago mundi* 45:77–89.

———. 1995. "From Books with Maps to Books as Maps: The Editor in the Creation of the Atlas," in Joan Winearls, ed., *Editing Early and Historical Atlases: Papers Given at the Twenty-ninth Annual Conference on Editorial Problems, University of Toronto, 5–6 November 1993*. Toronto: University of Toronto Press, pp. 3–48.

———, et al. 1993. *Two by Two: Twenty-two Pairs of Maps from the Newberry Library Illustrating 500 Years of Western Cartographic History*, by James Akerman, David Buisseret, and Robert W. Karrow, Jr. Chicago: The Newberry Library.

Andrews, J. H. 1969. Introduction to George Taylor and Andrew Skinner, *Maps of the Roads of Ireland*. Shannon: Irish University Press, pp. v–xvi.

———. 1997. "Friends in High Places: Gerard Mercator, 1564," in his *Shapes of Ireland: Maps and their Makers 1564–1839*. Dublin: Geography Publications, pp. 26–56.

Armstrong, Lilian. 1996. "Benedetto Bordon, *Miniator*, and Cartography in Early Sixteenth-Century Venice." *Imago mundi* 48:65–92.

Bechtel, Edwin De T. 1955. *Jacques Callot*. New York: George Braziller.

Bendall, A. Sarah. 1992. *Maps, Land and Society: A History, with a Carto-bibliography of Cambridgeshire Estate Maps, c. 1600–1836*. Cambridge: Cambridge University Press.

Berggren, J. Lennart, and Alexander Jones. 2000. *Ptolemy's Geography: An Annotated Translation of the Theoretical Chapters*. Princeton: Princeton University Press.

Bosse, David. 1982a. "Johan Gabriel Sack and his Maps," *Mapline* 26:1–4.

———. 1982b. "Pegging the River," *Mapline* 28:4–5.

Broecke, Marcel P. R. van den. 1996. *Ortelius Atlas Maps: An Illustrated Guide*. Houten, the Netherlands: HES Publishers.

———, et al. 1998. *Abraham Ortelius and the First Atlas: Essays Commemorating the Quadricentennial of his Death, 1598–1998*. Houten, the Netherlands: HES Publishers.

Brown, Lloyd A. 1949. *The Story of Maps*. Boston: Little, Brown.

———. 1952. *The World Encompassed*. Baltimore: Walters Art Gallery.

Buisseret, David. 1985. "A Cartographer's View of the French Empire in the Early 18th Century: The Collection Called *Cartes marines* at the Newberry Library, Chicago," in E. P. Fitzgerald, ed., *Proceedings of the Eighth Annual Meeting of the French Colonial Historical Society, 1982*. Lanham, Md.: University Press of America, pp. 9–11, pls. 1–3.

———. 1988. *Rural images: the estate plan in the Old and New Worlds. A cartographic exhibit at the Newberry Library. . . .* Chicago: The Newberry Library.

———. 1991. *Mapping the French Empire in North America*. Chicago: The Newberry Library.

———, ed. 1996. *Rural Images: Estate Maps in the Old and New Worlds*. Chicago: University of Chicago Press.

———, et al. 1998. *Envisioning the City: Six Studies in Urban Cartography*. Chicago: University of Chicago Press.

Campbell, Tony. 1968. *Claesz Jansz. Visscher: A Hundred Maps Described*. Map Collectors' Series, no. 46. London: The Map Collectors' Circle.

———. 1987. "Portolan Charts from the Late Thirteenth Century to 1500," in J. B. Harley and David Woodward, eds. *The History of Cartography*. Chicago: University of Chicago Press, vol. 1, pp. 371–463.

Cappon, Lester J. 1975. *Atlas of Early American History*. Princeton, N.J.: Princeton University Press.

Cavazzi, Lucia. 1982. *Luigi Rossini incisore: Vedute di Roma 1817–1850 [an exhibition at] Palazzo Braschi, 7 aprile – 15 luglio 1982*. Rome: Multigrafica Editrice.

Chase, George Wingate. 1861. *History of Haverhill, Massachusetts, from its First Settlement in 1640 to the Year 1860*. Haverhill, Mass.: Chase, pp. 217–31.

Cobarrubias, Gaspar de. 1971. *Relaciones de las Minas de Temascaltepeque y de los pueblos de Texcaltitlan, Cabecera de todos, Temazcaltepeque y Texupilco*, ed. by Javier Romero Quiroz. Toluca: Universidad Autónoma del Estado de México.

Conzen, Michael. 1984. "The County Landownership Map in America: Its Commercial Development and Social Transformation, 1814–1939," *Imago mundi* 36:9–31.

——. 1997. "The All-American County Atlas: Styles of Commercial Landownership Mapping and American Culture," in John A. Wolter and Ronald E. Grim, eds., *Images of the World: The Atlas through History*. Washington: Library of Congress, pp. 331–65.

Cook, Karen S. 2000. *Shedding the Light of Twentieth-Century Research on Dati's Sfera*. Research paper for History 500, "History of the Book," University of Kansas. Typescript.

Cortesão, Armando, and Avelino Teixeira da Mota. 1960–62. *Portugaliae monumenta cartographica*. Libson: Comisão Executiva das Comemorações do V Centenário da Morte do Infante D. Henrique. Reprinted at reduced size, Lisbon: Imprensa Nacional-Casa da Moeda, 1987.

Daniels, Stephen. 1999. *Humphry Repton: Landscape Gardening and the Geography of Georgian England*. New Haven : Published for the Paul Mellon Centre for Studies in the British Art [by] Yale University Press.

Davis, Gilbert A. 1874. *Centennial Celebration, together with an Historical Sketch of Reading, Windsor County, Vermont*. Bellows Falls, Vt.: A. N. Swain.

Dilke, O. A. W., *et al.* 1987. "The Culmination of Greek Cartography in Ptolemy," in J. B. Harley and David Woodward, eds., *The History of Cartography*. Chicago: University of Chicago Press, vol. 1, pp. 177–200.

Douville, Raymond. 1969. "Jean-Baptiste Hertel de Rouville," in *Dictionary of Canadian Biography*. Toronto: University of Toronto Press, vol. 2, pp. 284–86.

Evans, Geraint N. D. 1969. *Uncommon Obdurate: The Several Public Careers of J. F. W. DesBarres*. Salem, Mass.: Peabody Museum.

Evans, Ifor M., and Heather Lawrence. 1979. *Christopher Saxton, Elizabethan Map-maker*. Wakefield, Eng.: Wakefield Historical Publications.

Fernlund, Kevin J. 2000. *William Henry Holmes and the Rediscovery of the American West*. Albuquerque: University of New Mexico Press.

Fite, Emerson D., and Archibald Freeman. 1926. *A Book of Old Maps Delineating American History*. Cambridge: Harvard University Press, pp. 100–102.

Frassanito, William A. 1995. "Early Cartography and the Gettysburg Battlefield," in his *Early Photography at Gettysburg*. Gettysburg, Penn.: Thomas Publications, pp. 7–19.

Gardiner, R. A. 1971. "Edmond Halley's Isogonic Charts," *Geographic Journal* 137:419–20.

Gerhard, Peter. 1993. *Guide to the Historical Geography of New Spain*, 2d. ed., Norman: Univ. of Oklahoma Press, pp. 267–70.

Giono, Jean. 1965. *The Battle of Pavia, 24th February 1525*. London: Peter Owen.

Goetzman, William H. 1966. *Exploration and Empire: The Explorer and the Scientist in the Winning of the American West*. New York: Knopf.

Harley, J. B. 1990. *Maps and the Columbian Encounter: An Interpretive Guide to the Travelling Exhibition*. Milwaukee: Golda Meir Library, University of Wisconsin-Milwaukee.

——, *et al.* 1978. *Mapping the American Revolutionary War*, J. B. Harley, Barbara Bartz Petchenik, and Lawrence W. Towner. Chicago: University of Chicago Press.

Harley, J. B., and David Woodward, eds. 1987– . *The History of Cartography*. Chicago: University of Chicago Press.

Harris, Elizabeth. 1975. "Miscellaneous Map Printing Processes in the Nineteenth Century," in David Woodward, ed., *Five Centuries of Map Printing*. Chicago: University of Chicago Press, pp. 113–36.

Heidenreich, Conrad E. 1976. *Explorations and Mapping of Samuel de Champlain, 1603–1632*. Cartographica Monograph 17. Toronto: B. V. Gutsell.

Heijden, H. A. M. van der. 1992. *De oudste gedrukte kaarten van Europa*. Alphen aan den Rijn: Canaletto.

Hyde, Ralph. 1988. *Panoramania! The Art and Entertainment of the 'All-Embracing' View*. London: Trefoil Publications in association with Barbican Art Gallery.

Johnson, Richard Colles, and Cynthia H. Peters. 1986. *A Princely Gift: The Rudy Lamont Ruggles Collection of the Newberry Library*. Chicago: The Newberry Library.

Kain, R. J. P., and Elizabeth Baigent. 1992. *The Cadastral Map in the Service of the State: A History of Property Mapping*. Chicago: University of Chicago Press.

Karrow, Robert W. 1993. *Mapmakers of the Sixteenth Century and their Maps*. Winnetka, Ill.: Speculum Orbis Press for the Newberry Library.

Koeman, Cornelis. 1967–85. *Atlantes neerlandici: Bibliography of Terrestrial, Maritime, and Celestial Atlases and Pilot Books, Published in the Netherlands up to 1880*. Amsterdam: Theatrum Orbis Terrarum.

———. 1970. *Joan Blaeu and His Grand Atlas*. Amsterdam: Theatrum Orbis Terrarum.

Konvitz, Josef W. 1987. "The National Map Survey," in his *Cartography in France, 1660–1848: Science, Engineering, and Statecraft*. Chicago: University of Chicago Press, pp. 1–31.

Lambert, Audrey M. 1971. *The Making of the Dutch Landscape: An Historical Geography of the Netherlands*. London: Seminar Press.

Lanman, Jonathan T. 1983. "Folding or Collapsable Terrestrial Globes," *Der Globusfreund* 35/37:39–44.

———. 1987. *On the Origin of Portolan Charts*. The Hermon Dunlap Smith Center for the History of Cartography Occasional Publication no. 2. Chicago: The Newberry Library.

Laurie, Kedron. 2001. "Humphry Repton," *Grove Dictionary of Art*. Macmillan. 12 May 2001. <http://www.groveart.com>

Lyon. 2001. *Forma urbis: Les plans généraux de Lyon, XVIe–XXe siècles*. Archives de Lyon. 12 May 2001. <http://www.mairie-lyon.fr/fr/archives/fonds/plan-g/03.htm>

Martin, Lawrence. 1972. "John Disturnell's Map of the United Mexican States," in Walter W. Ristow, ed., *A la Carte: Selected Papers on Maps and Atlases*. Washington: Library of Congress, pp. 204–21.

Miller, Naomi. 2000. *Mapping Cities [an exhibition at] Boston University Art Gallery, January 14 – February 25, 2000*. Seattle: University of Washington Press.

Modelski, Andrew M. 1984. *Railroad Maps of North America: The First Hundred Years*. Washington: Library of Congress.

Morison, Samuel Eliot. 1972. *Samuel de Champlain, Father of New France*. Boston: Little, Brown.

Morrison, Walter K. 1988. "The Cartographical Revolution of 1775," in Barbara Farrell and Aileen Desbarats, eds., *Explorations in the History of Canadian Mapping: A Collection of Essays*. Ottawa: Association of Canadian Map Libraries and Archives, pp. 75–88.

Mundy, Barbara E. 1996. *The Mapping of New Spain: Indigenous Cartography and the Maps of the Relaciones Geográficas*. Chicago : University of Chicago Press.

———. 1998. "Mapping the Aztec Capital: The 1524 Nuremberg Map of Tenochtitlan, Its Sources and Meanings," *Imago mundi* 50:11–33.

Nebenzahl, Kenneth. 1974. *Atlas of the American Revolution*. Chicago: Rand McNally & Co.

———. 1975. *A Bibliography of Printed Battle Maps of the American Revolution, 1775–1795*. Chicago: University of Chicago Press.

Norwich, Oscar I. 1983. *Maps of Africa: An Illustrated and Annotated Carto-bibliography*. Johannesburg: Ad. Donker.

Osley, A. S. 1969. *Mercator: A Monograph on the Lettering of Maps, etc. in the 16th Century Netherlands. With a Facsimile and Translation of his Treatise on the Italic Hand and a Translation of Ghim's Vita Mercatoris*. New York: Watson-Guptill.

Parke-Bernet. 1969. *The Original Manuscript Map of the Military District, Kansas and the Territories (1866) by Maj. Gen. Grenville Mellen Dodge*. Sale no. 2934A. New York: Parke-Bernet Galleries.

Pastoureau, Mireille. 1984. *Les atlas français XVIe–XVIIe siècles*. Paris: Bibliothèque Nationale.

———, ed. 1988. *Nicolas Sanson, Atlas du monde 1665*. Paris: Sand & Conti.

Pedley, Mary. 1981. "The Map Trade in Paris, 1650–1825," *Imago mundi* 33:33–45.

Pelletier, Monique. 1990. *La carte de Cassini: L'extraordinaire aventure de la carte de France*. Paris: Presses de l'École Nationale des Ponts et Chaussées.

Peters, Cynthia. 1984. "Rand, McNally in the Nineteenth Century: Reaching for a National Market," in Michael P. Conzen, ed., *Chicago Mapmakers: Essays on the Rise of the City's Map Trade*. Chicago: Chicago Historical Society for the Chicago Map Society, pp. 64–72.

Pollak, Martha. 1991. *Military Architecture, Cartography, and the Representation of the Early Modern European City: A Checklist of Treatises on Fortification in the Newberry Library*. Chicago: The Newberry Library, pp. 54–55.

Preuss, Charles. 1958. *Exploring With Frémont: The Private Diaries of Charles Preuss, Cartographer for John C. Frémont on his First, Second, and Fourth Expeditions to the Far West*. Norman: University of Oklahoma Press.

Richardson, John. 2000. "Pál Teleki, A Hungarian Cartographer in Paris," *Mapline* 91:1–4.

Ristow, Walter W. 1946. "American Road Maps and Guides," *Scientific Monthly* (May):397–406.

————. 1961. *A Survey of the Roads of the United States of America, 1789*. Cambridge: The Belknap Press of Harvard University Press.

————. 1972. "John Mitchell's Map of the British and French Dominions in North America," in his *A la Carte: Selected Papers on Maps and Atlases*. Washington: Library of Congress, pp. 102–13.

————. 1974. "Dutch Polder Maps," *Quarterly Journal of the Library of Congress* 31:136–49.

————. 1977. "Robert Mills' Atlas of the State of South Carolina, 1825: The First American State Atlas." *Quarterly Journal of the Library of Congress* 34:52–56.

————. 1985. *American Maps and Mapmakers: Commercial Cartography in the Nineteenth Century*. Detroit: Wayne State University Press.

————. 1997. "Early American Atlases and Their Publishers," in John A. Wolter and Ronald E. Grim, *Images of the World: The Atlas through History*. Washington: Library of Congress, pp. 301-29.

Robinson, Arthur H. 1982. *Early Thematic Mapping in the History of Cartography*. Chicago: University of Chicago Press.

Ross, M. J. 1982. *Ross in the Antarctic: The Voyages of James Clark Ross in Her Majesty's Ships* Erebus & Terror, *1839–1843*. Whitby, Eng.: Caedmon.

Rouleau, Bernard. 1989. *Le Plan de Paris de Louis Bretez dit Plan de Turgot, présenté et commenté par Bernard Rouleau*. Nördlingen, Germany: Verlag Dr. Alfons Uhl.

Russ, C. J. 1974. "Jean-Baptiste de Saint-Ours Deschaillons," in *Dictionary of Canadian Biography*. Toronto: University of Toronto Press, vol. 3, pp. 578–79.

Russell, H. Diane. 2001. "Jacques Callot," *Grove Dictionary of Art*. Macmillan. 12 May 2001. <http://www.groveart.com>

Savory, Reginald. 1966. *His Britannic Majesty's Army in Germany during the Seven Years War*. Oxford: Clarendon Press, pp. 25–46.

Schulz, Juergen. 1970. *The Printed Plans and Panoramic Views of Venice (1486–1797)*. Saggi e memorie di storia dell'arte, 7. Florence: Leo S. Olschki.

————. 1978. "Jacopo de' Barbari's View of Venice: Map Making, City Views, and Moralized Geography before the Year 1500," *Art Bulletin* 60:425–74.

Schwartz, Seymour. 1994. *The French and Indian War, 1754–1763*. New York: Simon & Schuster.

Shefrin, Jill. 1999. *Neatly Dissected for the Instruction of Young Ladies and Gentlemen in the Knowledge of Geography: John Spilsbury and Early Dissected Puzzles*. Los Angeles: Cotsen Occasional Press.

Shirley, Rodney W. 1993. *The Mapping of the World: Early Printed World Maps, 1472–1700*. London: New Holland.

Sinistri, Cesare, and Luigi Casali. 1996. "Il ritrovamento dell piu antica pianta di Pavia," *Bollettino della Societá Pavese di Storia Patria*, n.s. 48:481–84.

Skelton, R. A. 1963. Bibliographical note in *Claudius Ptolemaeus, Cosmographia, Ulm 1482*. Amsterdam: Theatrum Orbis Terrarum, pp. v–[xii].

————. 1965–66. Introduction in Georg Braun and Frans Hogenberg, *Civitates orbis terrarum 'The Towns of the World'*. Amsterdam: Theatrum Orbis Terrarum, pp. vii–xlvii.

————. 1966a. Bibliographical note in *Benedetto Bordone, Libro . . . de tutte l'isole del mondo, Venice 1528*. Amsterdam: Theatrum Orbis Terrarum, pp. v–xii.

———. 1966b. Bibliographical note in *Francesco Berlinghieri, Geographia, Florence 1482*. Amsterdam: Theatrum Orbis Terrarum, pp. v–xiii.

Smith, Clara E. 1927. *List of Manuscript Maps in the Edward E. Ayer Collection*. Chicago: The Newberry Library.

Snyder, John P. 1993. *Flattening the Earth: Two Thousand Years of Map Projections*. Chicago: University of Chicago Press.

Stevens, Henry Newton. 1937. *Catalogue of the Henry Newton Stevens Collection of the Atlantic Neptune, Together with a Concise Bibliographical Description of Every Chart, View and Leaf of Text Contained therein, as also of Certain Other States Observed Elsewhere*. London: H. Stevens, Son & Stiles.

Storm, Colton. 1968. *A Catalogue of the Everett D. Graff Collection of Western Americana*. Chicago: For the Newberry Library by the University of Chicago Press.

Szczesniak, Boleslaw. 1956. "The Seventeenth Century Maps of China: An Inquiry into the Compilations of European Cartographers," *Imago mundi* 13:116–36.

Talbot, Charles. 1982. "Topography as Landscape in Early Printed Books," in Sandra Hindman, ed., *The Early Illustrated Book: Essays in Honor of Lessing J. Rosenwald*. Washington: Library of Congress, pp. 105–16.

Thrower, Norman W. 1981. *The Three Voyages of Edmond Halley in the Paramore, 1698–1701*. Hakluyt Society, Works, vols. 156–57. London: The Hakluyt Society.

Thurman, Mel. 1989. "Warren, Dodge, and Later Nineteenth-Century Army Maps of the West," *Mapline* 63:1–4.

Tyacke, Sarah, and John Huddy. 1980. *Christopher Saxton and Tudor Map-making*. London: British Library.

Unruh, John D., Jr. 1979. *The Plains Across*. Urbana: University of Illinois Press.

Virginia. 1995. "Exploring the West from Monticello: Exhibition of Maps and Navigational Instruments, on View in the Tracy W. McGregor Room, Alderman Library, University of Virginia, July 10 to September 26, 1995." University of Virginia. 12 May 2001. <http://www.lib.virginia.edu/exhibits/lewis_clark/>

Wagner, Henry R. 1931. "The Manuscript Atlases of Battista Agnese," *Papers of the Bibliographical Society of America* 25:1–110.

———. 1947. "Additions to the Manuscript Atlases of Battista Agnese," *Imago mundi* 4:28–30.

Watelet, Marcel, ed. 1998. *The Mercator Atlas of Europe: Facsimile of the Maps by Gerardus Mercator Contained in the Atlas of Europe, Circa 1570–1572*. Pleasant Hill, Ore.: Walking Tree Press.

Wheat, Carl Irving. 1957–63. *Mapping the Transmississippi West, 1540–1861*. San Francisco: Institute of Historical Cartography.

Wilson, Samuel, Jr. 1982. "French Fortification at Fort Rosalie, Natchez," in Patricia K. Galloway, ed., *La Salle and His Legacy: Frenchmen and Indians in the Lower Mississippi Valley*. Jackson: University Press of Mississippi, pp. 194–210.

Wolff, Hans, ed. 1989. *Philipp Apian und die Kartographie der Renaissance*. Weissenhorn, Germany: Anton H. Konrad Verlag.

Woodward, David. 1977. *The All-American Map: Wax Engraving and its Influence on Cartography*. Chicago: University of Chicago Press.

———. 1996. *Maps as Prints in the Italian Renaissance: Makers, Distributors & Consumers*. Panizzi Lectures, 1995. London: British Library.

WPA. 1976. *A History of Spartanburg County, Compiled by the Work Projects Administration in the State of South Carolina*. American Guide Series. Spartanburg, S.C.: Reprint Co.